my **revision** notes

OCR GCSE (9–1)

GEOGRAPHY B

Rebecca Blackshaw

Jo Payne

Simon Ross

 **HODDER**
EDUCATION
AN HACHETTE UK COMPANY

The Publishers would like to thank the following for permission to reproduce copyright material.

Photo credits p.11 *l* © Rex Wholster – stock.adobe.com; **p.11** *r* © National Geographic Creative / Alamy Stock Photo; **p.20** © Shutterstock / Amy Laughinghouse; **p.52** *t* © Ian Ward; **p.52** *b* © Robert Estall photo agency / Alamy Stock Photo; **p.55** © Graham M. Lawrence / Alamy Stock Photo; **p.65** © RSM Images / Alamy Stock Photo; **p.66** © Mike Goldwater / Alamy Stock Photo; **p.72** © John Gaps III/AP/REX/Shutterstock; **p.81** © Joerg Boethling / Alamy Stock Photo; **p.91** © World Mapper; **p.127** © pixelbliss/123RF.com; **p.130** © S.R.Miller – stock.adobe.com; **p.141** © Carolyn Cole/Los Angeles Times via Getty Images

Acknowledgements
Every effort has been made to trace all copyright holders, but if any have been inadvertently overlooked, the Publishers will be pleased to make the necessary arrangements at the first opportunity.

Although every effort has been made to ensure that website addresses are correct at time of going to press, Hodder Education cannot be held responsible for the content of any website mentioned in this book. It is sometimes possible to find a relocated web page by typing in the address of the home page for a website in the URL window of your browser.

Hachette UK's policy is to use papers that are natural, renewable and recyclable products and made from wood grown in sustainable forests. The logging and manufacturing processes are expected to conform to the environmental regulations of the country of origin.

Orders: please contact Bookpoint Ltd, 130 Park Drive, Milton Park, Abingdon, Oxon OX14 4SE. Telephone: +44 (0)1235 827720. Fax: +44 (0)1235 400401. Email education@bookpoint.co.uk Lines are open from 9 a.m. to 5 p.m., Monday to Saturday, with a 24-hour message answering service. You can also order through our website: www.hoddereducation.co.uk

ISBN: 978 1 4718 8734 5

First published in 2017 by
Hodder Education,
An Hachette UK Company
Carmelite House
50 Victoria Embankment
London EC4Y 0DZ
www.hoddereducation.co.uk

Impression number   10 9 8 7 6 5 4 3 2

Year   2021 2020 2019 2018

Cover photo © Igor Dmitriev/123RF.com
Illustrations by Integra Software Services Pvt. Ltd. and Barking Dog Art
Typeset in Integra Software Services Pvt. Ltd., Pondicherry, India
Printed in Spain

A catalogue record for this title is available from the British Library.

Get the most from this book

Everyone has to decide his or her own revision strategy, but it is essential to review your work, learn it and test your understanding. These Revision Notes will help you to do that in a planned way, topic by topic. Use this book as the cornerstone of your revision and don't hesitate to write in it – personalise your notes and check your progress by ticking off each section as you revise.

Tick to track your progress

Use the revision planner on pages 4–7 to plan your revision, topic by topic. Tick each box when you have:
- revised and understood a topic
- tested yourself
- practised the exam questions and gone online to check your answers.

You can also keep track of your revision by ticking off each topic heading in the book. You may find it helpful to add your own notes as you work through each topic.

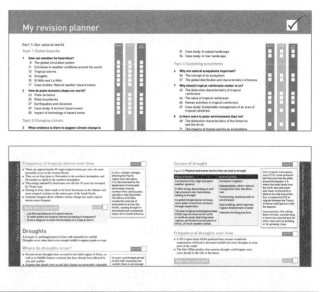

Features to help you succeed

Exam tips

Expert tips are given throughout the book to help you polish your exam technique in order to maximise your chances in the exam.

Now test yourself

These short, knowledge-based questions provide the first step in testing your learning. Answers can be found online at: **www.hoddereducation.co.uk/myrevisionnotes**

Definitions and key words

Clear, concise definitions of essential key terms are provided where they first appear.

Revision activities

These activities will help you to understand each topic in an interactive way.

Exam practice

Practice exam questions are provided for each topic. Use them to consolidate your revision and practise your exam skills.

Online

Go online to check your answers to the exam questions at **www.hoddereducation.co.uk/myrevisionnotes**

My revision planner

Part 1: Our natural world

Topic 1 Global hazards

1 How can weather be hazardous?
- 8 The global circulation system
- 11 Extremes in weather conditions around the world
- 13 Tropical storms
- 14 Droughts
- 15 El Niño and La Niña
- 17 Case studies: Natural weather hazard events

2 How do plate tectonics shape our world?
- 23 Plate tectonics
- 25 Plate boundaries
- 27 Earthquakes and volcanoes
- 29 Case study: A tectonic hazard event
- 30 Impact of technology in hazard zones

Topic 2 Changing climate

3 What evidence is there to suggest climate change is a natural process?
- 31 The pattern of climate change
- 32 Evidence for climate change
- 33 Causes of natural climate change
- 34 The natural greenhouse effect
- 36 Impacts of climate change worldwide
- 37 Impacts of climate change within the UK

Topic 3 Distinctive landscapes

4 What makes a landscape distinctive?
- 38 The concept of a landscape
- 39 The distribution of landscapes in the UK
- 40 The characteristics of landscapes in the UK

5 What influences the landscapes of the UK?
- 42 Geomorphic processes
- 45 Coastal landforms
- 47 River landforms

REVISED   TESTED   EXAM READY

51   Case study: A coastal landscape
54   Case study: A river landscape

Topic 4 Sustaining ecosystems

6   Why are natural ecosystems important?
56   The concept of an ecosystem
57   The global distribution and characteristics of biomes

7   Why should tropical rainforests matter to us?
62   The distinctive characteristics of tropical rainforests
64   The value of tropical rainforests
65   Human activities in tropical rainforests
67   Case study: Sustainable management of an area of tropical rainforest

8   Is there more to polar environments than ice?
68   The distinctive characteristics of the Antarctic and the Arctic
71   The impacts of human activity on ecosystems in the Antarctic and the Arctic
75   Case study: Small-scale sustainable management in the Antarctic and the Arctic
76   Case study: Global sustainable management in the Antarctic and the Arctic

Part 2: People and society

Topic 5 Urban futures

9   Why do more than half the world's population live in urban areas?
77   The global pattern of urbanisation
78   World cities and megacities
80   Rapid urbanisation in cities
82   Urban trends in advanced countries

10   What are the challenges and opportunities for cities today?
84   Case study: A city in an advanced country
86   Case study: A city in an emerging developing country

REVISED   TESTED   EXAM READY

Topic 6 Dynamic development

REVISED   TESTED   EXAM READY

11 Why are some countries richer than others?

88 The definition of development and how countries are classified

89 How development is measured

92 Factors influencing uneven development

94 Factors that make it hard for countries to break out of poverty

12 Are LIDCs likely to stay poor?

95 Economic development in Ethiopia

96 A model of economic development

97 Millenium Development Goals

98 Wider political, social and environmental factors affecting Ethiopia's development

99 International trade

101 The role of TNCs

102 Aid and debt relief

104 Development strategies

Topic 7 The UK in the 21st century

13 How is the UK changing in the 21st century?

106 The characteristics of the UK

110 Population trends and the Demographic Transition Model (DTM)

112 The UK's ageing population

113 Population change in a named place: Boston, Lincolnshire

115 Economic changes in the UK

117 UK economic hubs

118 The changes in one economic hub

14 Is the UK losing its global significance?

120 The UK's role in political conflict

120 Case study: The UK's political role in one global conflict

122 The UK's media exports

124 Contribution of ethnic groups to the cultural life of the UK

Topic 8 Resource reliance

15 Will we run out of natural resources?

125 Supply and demand of food, water and energy

127 How environments and ecosystems are modified by humans

16 Can we feed nine billion people by 2050?

131 Food security and the factors which influence it

133 World patterns of food access

134 Malthusian and Boserupian theories

135 Case study: Food security

138 Sustainable strategies to achieve food security

Part 3: Fieldwork and geographical exploration

17 143 Geography fieldwork

18 144 Geographical exploration

Now test yourself and exam practice answers at www.hoddereducation.co.uk/myrevisionnotes

1 How can weather be hazardous?

The global circulation system

How does it work?

REVISED

There are three large-scale circular movements of air in each **hemisphere** of the Earth's surface. These circular movements, or 'cells', take air from the Equator and move it towards the poles. The cells have a role to play in creating the **climate zones** on Earth.

> **Hemisphere:** one half of the Earth, usually divided into northern and southern halves by the Equator.
>
> **Climate zone:** divisions of the Earth's climates into belts, or zones, according to average temperatures and average rainfall. The three major zones are polar, temperate and tropical.

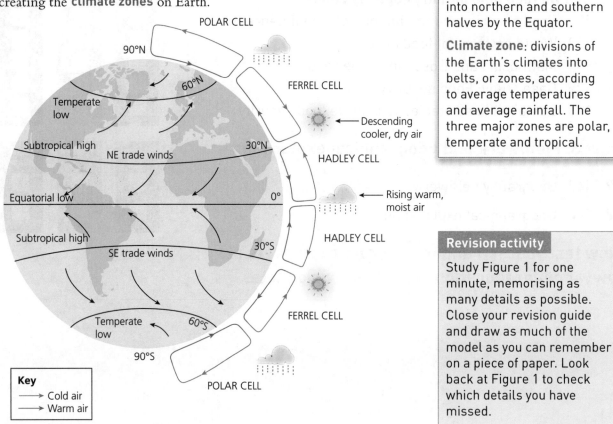

Figure 1 The global circulation system

> **Revision activity**
>
> Study Figure 1 for one minute, memorising as many details as possible. Close your revision guide and draw as much of the model as you can remember on a piece of paper. Look back at Figure 1 to check which details you have missed.

Circulatory cells

REVISED

Figure 2 Characteristics of the circulatory cells

	Where is it?	What happens?
Hadley cell	The largest cell, which extends from the Equator to 30° in the north and south.	Winds meet near the Equator and the warm air rises, causing thunderstorms. The drier air then flows out towards 30° before sinking over subtropical areas.
Ferrel cell	The middle cell, which generally occurs from the edge of the Hadley cell at 30° to 60° in the north and south.	Air in this cell joins the sinking air at the edge of the Hadley cell; it travels across these mid-latitude regions until the air rises along the border of cold air with the Polar cell.
Polar cell	The smallest and weakest cell, which occurs from the edge of the Ferrel cell to the poles at 90°.	The air sinks over the higher latitudes at the poles and flows towards the mid-latitudes where it meets the Ferrel cell and rises.

High and low pressure

Atmospheric air pressure ranges from low pressure of approximately 980 millibars to high pressure of approximately 1050 millibars. **Low pressure** is created where the two Hadley cells meet and air rises. Where the Hadley and Ferrel cells meet, air descends creating **high pressure**.

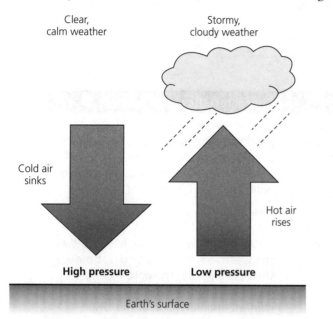

Figure 3 High and low pressure

> **Atmospheric air pressure**: the force exerted on the Earth's surface by the weight of the air, measured in millibars.
>
> **Low pressure**: occurs when the air is rising, so less air is pressing down on the ground; air rises as it warms, leading to low pressure at the surface.
>
> **High pressure** occurs when there is more air pressing down on the ground caused by air sinking; air descends as it cools, leading to high pressure at the surface.
>
> **Precipitation**: the collective term for moisture that falls from the atmosphere; it could be in the form of rain, sleet, snow or hail.

High pressure

- When air cools it becomes denser and falls towards the ground, leading to high pressure.
- Cool air warms as it reaches the Earth's surface, causing any clouds to evaporate.
- Heavy rain at the Equator means that most of the moisture has gone by the time the air reaches the subtropics.
- High-pressure systems are usually associated with clear skies and dry, hot weather.

Low pressure

- Low pressure causes warm air to rise, after which it cools and condenses to form clouds.
- Moisture falls from the atmosphere as rain, sleet, snow or hail (collectively known as **precipitation**).
- Differences in temperature between day and night are unlikely to be large as the cloud cover reflects solar radiation during the day and traps it at night.

Now test yourself

TESTED

1 At which line of latitude do the Polar and Ferrel cells meet?
2 Would an air pressure of 1036 millibars be high or low pressure?
3 Which circulatory air cell is the smallest?
4 Which circulatory cells meet at the Equator?
5 Why might the climatic conditions be unsettled around 60° latitude in the northern and southern hemispheres?

Climate zones

Exam tip

The specification requires you to be able to describe the relationship between the global circulation system and the climate zones. Make sure you refer to precise details such as the cells, continents/regions, air pressure and latitude.

Figure 4 Climate zones

Climate zone	Latitude	Characteristics of the climate
Polar	At the poles 90° north and south of the Equator	Cold air from the Polar cell sinks, producing high pressure. The spin of the Earth creates dry, icy winds. In some parts of Antarctica, the average wind speed is 80 kph.
Temperate	Mid-latitudes 50° to 60° north and south of the Equator	Two air cells meet, one warm from the Ferrel cell and one cold from the Polar cell. Low pressure is created as the warm air from the Equator meets the cold air from the poles along a weather front. This brings frequent rainfall and is typical of the UK.
Subtropical	30° north and south of the Equator	High pressure as a result of sinking air where Hadley and Ferrel cells meet. This creates a belt of deserts, including the Sahara in northern Africa and the Namib in southern Africa. Daytime temperatures can exceed 40°C.
Tropical	At the Equator, 0° line of latitude	A belt of low pressure where the Hadley cells meet and air rises rapidly. This results in regular heavy rainfall and thunderstorms in places such as Malaysia in South East Asia and northern Brazil in South America.

Now test yourself

1 Which climate zone is found where Hadley and Ferrel cells meet?
2 Brazil and Malaysia are examples of which climate zone?
3 Why do deserts form at 30° north and south of the Equator?

Exam practice

1 What is the global circulation system? [2]
2 Describe the climatic conditions in a high-pressure belt. [2]
3 State a link between the Hadley cells and tropical climates. [2]

ONLINE ☐

Extremes in weather conditions around the world

Temperature

REVISED

Coldest place

- **Vostok, Antarctica**: on 21 July 1983, the coldest air temperature ever was recorded at the Russian research station, Vostok: −89.2 °C. It has an altitude of around 3500 m, which helps to make it the coldest place on Earth. For every 1 km in altitude the temperature decreased by 6.5 °C.

Hottest place

- **Al-Aziziyah, Libya**: on 13 September 1922, the world experienced its hottest air temperature ever recorded at 57.8 °C in Libya, which is located 32° north of the Equator. This means Libya is in the subtropical high region.

Precipitation

REVISED

Driest places

- **Death Valley, USA**: one of the driest places in North America with an average rainfall of 60 mm per year. Storms from the Pacific Ocean travel over a series of mountain ranges before they reach Death Valley, meaning that the moisture has already fallen as rain (Figure 5).
- **Aswan, Egypt**: it has an average rainfall of only 0.861 mm per year; it is close to the Tropic of Cancer.
- **Atacama Desert, South America**: the average annual rainfall is 15 mm. This is due to its location in the rain shadow of the Andes. On its western side, the onshore winds do not have enough warmth to pick up moisture from the ocean surface.

Wettest places

- **Mawsynram, India**: this village of 10,000 people copes with an annual average rainfall of 11,871 mm, 80 per cent of which arrives during the seasonal **monsoon** (Figure 6).
- **Ureca, Equatorial Guinea**: located on the southern tip of Bioko Island, Ureca is the wettest place in Africa, with 10,450 mm per year.

Figure 6 A sign declaring Mawsynram as the wettest place on Earth

> **Monsoon**: heavy rain that arrives as a result of seasonal wind, most notably in southern Asia and India between May and September.

Figure 5 Death Valley, USA

Windiest places

- **Commonwealth Bay, Antarctica**: winds regularly exceed 240 kilometres per hour, with an average annual wind speed of 80 kilometres per hour. Winds carry air from high ground down the slopes by gravity.
- **Wellington, New Zealand**: the strongest gust of wind recorded in Wellington was 248 kilometres per hour. Gusts of wind exceed gale force on 175 days of the year. The mountains either side of Wellington funnel the winds.

Exam tip

You will be required to use your knowledge of specific places and facts to support your argument. Make sure you are familiar with the map of the world.

Now test yourself

TESTED

1 Where is the hottest place on Earth and what temperature was measured there?
2 Why is rainfall low in the Atacama Desert?
3 How are the winds in Wellington, New Zealand, intensified?

Exam practice

The table below lists a selection of the wettest places in the world and their annual rainfall totals.

Location	Rainfall (mm)
Big Bog in Maui, Hawaii	10,272
Debundscha, Cameroon	10,229
Mawsynram, India	11,777
Mount Emei, China	8,169
River Cropp waterfall, New Zealand	11,516
Tutendo, Colombia	11,770

1 Use the table to calculate the:
  (a) median
  (b) mean
  (c) range of the data.
2 Suggest an appropriate graphical technique to present this data.

ONLINE

Revision activity

Use a blank map of the world to locate the extremes of weather and annotate it with the important information. Do you notice a pattern? Can you link it to your knowledge of the global circulation model and high/low pressure? Are there any anomalies?

Tropical storms

Tropical storms begin as low-pressure systems in the tropics. They develop into tropical cyclones (also known as hurricanes or typhoons depending on their geographical location) when wind speeds reach 119 kilometres per hour.

Where do tropical storms occur?

REVISED

Tropical storms only happen in certain areas:
- typically between 5° and 15° north and south of the Equator
- temperature of the surface of the ocean more than 26.5 °C
- ocean depth of at least 50–60 m
- At least 500 kilometres away from the Equator so that the **Coriolis effect** can make the weather system rotate.

> **Tropical storm:** an area of low pressure with winds moving in a spiral around a calm central point called the 'eye' of the storm. The winds are powerful and rainfall is heavy.

> **Coriolis effect:** the effect of the Earth's rotation on weather patterns and ocean currents, making storms swirl clockwise in the southern hemisphere and anticlockwise in the northern hemisphere.

Exam tip

It is easy to lose marks by giving oversimplified statements, for example stating that 'the ocean water has to be warm' instead of that 'the sea surface temperature has to be at least 26.5 °C'. Be as precise as you can in your descriptions.

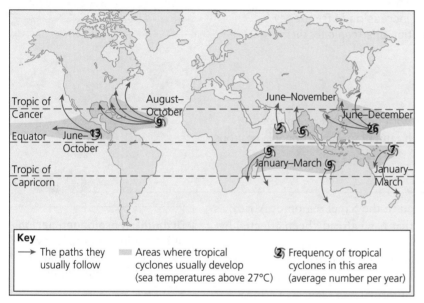

Key
| → | The paths they usually follow | | Areas where tropical cyclones usually develop (sea temperatures above 27°C) | ⑤ | Frequency of tropical cyclones in this area (average number per year) |

Figure 7 Global distribution and frequency of tropical storms

Causes of tropical storms

REVISED

A number of factors contribute to the development of a tropical storm:
- Temperatures need to cool quickly enough for tall clouds to form through condensation
- The wind speeds need to change slowly with height – this is known as wind shear; if the winds in the upper and lower atmosphere are different speeds, the storm will be torn apart
- Fuelled by warm ocean water, water vapour is rapidly drawn upwards into the low-pressure system; deep clouds rise from the Earth's surface to 15 km
- The most destruction occurs at the eyewall where the wind speeds are greatest and rainfall heaviest; this is typically 15–30 km from the centre of the storm
- When vertical winds reach the top of the troposphere at 16 km, they are deflected outwards by the Coriolis effect; this is what makes the storm rotate.

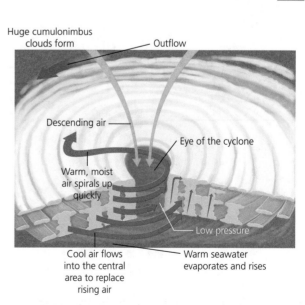

Figure 8 Cross section of a tropical storm

Frequency of tropical storms over time

- There are approximately 80 major tropical storms per year; the most powerful occur in the western Pacific.
- They occur from June to November in the northern hemisphere and November to April in the southern hemisphere.
- The energy released by hurricanes over the last 30 years has increased by 70 per cent.
- During **El Niño**, there tends to be fewer hurricanes in the Atlantic and more tropical cyclones in the eastern part of the South Pacific.
- Scientists disagree about whether climate change has made tropical storms more frequent.

> **El Niño**: climatic changes affecting the Pacific region every few years. It is characterised by the appearance of unusually warm water around northern Peru and Ecuador, typically in late December. The effects of El Niño include the reversal of wind patterns across the Pacific, causing drought in Australasia and unseasonal heavy rain in South America.

Now test yourself

TESTED

1 List five key features of tropical storms.
2 To what extent are tropical storms increasing in frequency?
3 Draw a diagram to show the formation of a tropical storm.

Droughts

A **drought** is a prolonged period of time with unusually low rainfall. Droughts occur when there is not enough rainfall to support people or crops.

Where do droughts occur?

- Recent severe droughts have occurred in the Sahel region of Africa, as well as in Middle Eastern countries that have already been affected by war and conflict.
- Regions that already have an arid (dry) climate are particularly vulnerable if they receive less than their usually very low rainfall. These include Australia, parts of the USA (such as California) and regions of China.
- There are some unexpected examples of drought in the world, such as the Amazon Basin in Brazil, where a drought affected 19 million square kilometres of rainforest between 2002 and 2005.

> **Drought**: a prolonged period of time with unusually low rainfall; there is not enough rainfall to support people or crops.

> **Exam tip**
>
> The definition of a drought continues to pose a challenge for students. It does NOT mean there is no rainfall. The rainfall is less than usual for a period of time. This is how places like the Amazon can experience drought.

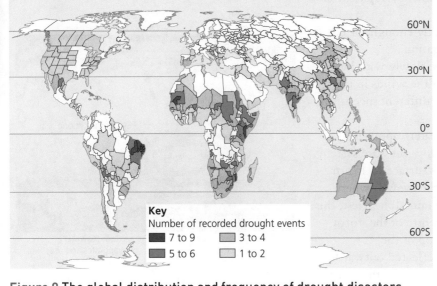

Figure 9 The global distribution and frequency of drought disasters, 1974–2004

Causes of drought

Figure 10 Physical and human factors that can lead to drought

Physical factors	Human factors
A presence of dry, high-pressure weather systems	Excessive irrigation
El Niño brings descending air and high pressure over Australasia, leading to drought	**Deforestation**, which reduces transpiration and, therefore, rain
As global temperatures increase, more water is lost from surfaces through evaporation	Overgrazing, exposing soils to wind erosion
The **inter-tropical convergence zone (ITCZ)** may not move as far north or south as usual, depriving some regions, particularly across parts of Africa, of much-needed rainfall	Dam building, which deprives regions downstream of water
	Intensive farming practices

Inter-tropical convergence zone (ITCZ): a low-pressure belt that encircles the globe around the Equator. It is where the trade winds from the north-east and south-east meet. As the Earth is tilted on its orbit around the Sun, it causes the ITCZ to migrate between the Tropics of Cancer and Capricorn with the seasons.

Deforestation: the cutting down of trees, transforming a forest into cleared land for other uses such as building, or for growing crops.

Frequency of droughts over time

- A 2013 report from NASA predicted that warmer worldwide temperatures will lead to decreased rainfall and more droughts in some parts of the world.
- The Met Office predicts that extreme drought could happen once every decade in the UK in the future.

Now test yourself

1 Explain how human factors can make the effects of a drought worse.
2 What is the ITCZ and how can it cause a drought?

El Niño and La Niña

What causes El Niño?

Scientists continue to study the causes of El Niño. It was once thought that sea floor heating following volcanic activity caused it, but this theory is unlikely. Small changes in sea surface temperatures are a more probable cause, possibly from tropical storms, which trigger the movement of water in a different direction.

Extreme weather conditions

Figure 11 compares normal conditions with the conditions in years when
El Niño and La Niña occur.

Figure 11 Comparison of normal weather conditions with El Niño and La Niña years

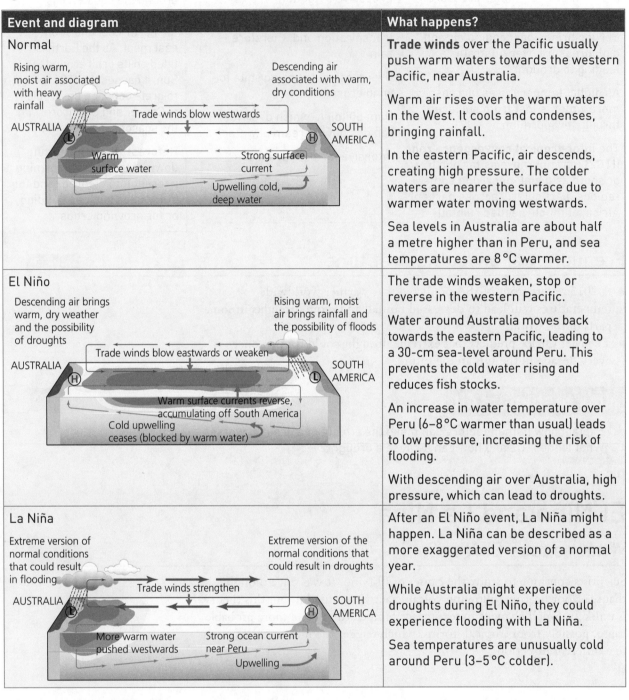

Event and diagram	What happens?
Normal	**Trade winds** over the Pacific usually push warm waters towards the western Pacific, near Australia. Warm air rises over the warm waters in the West. It cools and condenses, bringing rainfall. In the eastern Pacific, air descends, creating high pressure. The colder waters are nearer the surface due to warmer water moving westwards. Sea levels in Australia are about half a metre higher than in Peru, and sea temperatures are 8°C warmer.
El Niño	The trade winds weaken, stop or reverse in the western Pacific. Water around Australia moves back towards the eastern Pacific, leading to a 30-cm sea-level around Peru. This prevents the cold water rising and reduces fish stocks. An increase in water temperature over Peru (6–8°C warmer than usual) leads to low pressure, increasing the risk of flooding. With descending air over Australia, high pressure, which can lead to droughts.
La Niña	After an El Niño event, La Niña might happen. La Niña can be described as a more exaggerated version of a normal year. While Australia might experience droughts during El Niño, they could experience flooding with La Niña. Sea temperatures are unusually cold around Peru (3–5°C colder).

Exam practice

1 Describe the global distribution of tropical storms. [3]
2 Outline the conditions needed for a tropical storm to form. [3]
3 Explain two physical causes of drought. [4]

ONLINE

Trade winds: the prevailing pattern of easterly surface winds found in the tropics, within the lower section of the Earth's atmosphere.

Case studies: Natural weather hazard events

Exam tip

You need to know two case studies of natural weather hazard events. You have the choice of flash flooding or tropical storms for one, and a heatwave or a drought for the other. You must make sure that one is **UK based** and the other **non-UK based**.

Case study: Drought in Australia

Background

From 2002 to 2009, Australia experienced its worst drought for 125 years, which became known as the Big Dry.

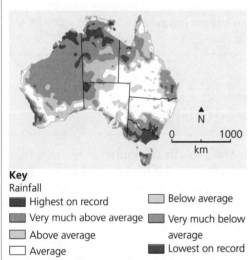

Key
Rainfall
- ■ Highest on record
- ■ Very much above average
- ▨ Above average
- □ Average
- ▨ Below average
- ■ Very much below average
- ■ Lowest on record

Figure 12 The distribution of rainfall in Australia, 1997–2009

Causes

The fact that Australia is often affected by droughts is influenced by a number of factors:

- Australia's geographical location makes it vulnerable to droughts. It is in a subtropical area of the world that experiences dry, sinking air leading to clear skies and little rain.
- In 2006, the rainfall was 40–60 per cent below normal over most of Australia south of the Tropic of Capricorn.

- When El Niño is in action, the chances of rainfall in Australia decrease and it becomes even drier than normal, particularly in eastern Australia.
- The Murray–Darling river basin is home to 2 million people and is under a lot of pressure to supply water to residents and for agricultural production.

Consequences

Figure 13 shows the devastating consequences of the Big Dry.

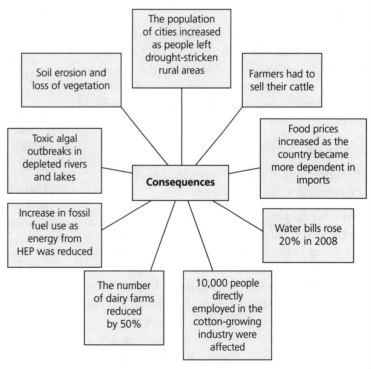

Figure 13 Consequences of the Big Dry

Responses

The different stakeholders found responses to the Big Dry.

Figure 14 Responses to the Big Dry

Individuals	Recycling waste water from showers, baths and wash basins (grey water)
	Farmers claiming financial assistance of $400–600 per fortnight
Local government	Subsidising rainwater storage tanks for homes
	Legislation to ban car washing and limit showers to four minutes
National government	A new multimillion-dollar desalinisation plant built in Sydney
	Paying out $1.7 million a day in drought relief to farmers
Scientists and environmentalists	More efficient irrigation systems
	Calculating the amount of water that can sustainably be used by a state to create a limit that could be traded across states

1 Describe one social, one economic and one environmental consequence of drought in Australia.
2 Choose two responses at different scales from Figure 14. Explain how they would help to reduce the effects of the drought.

Revision activity

Copy the spider diagram in Figure 13. Use three different colours to indicate the social, economic and environmental consequences. Add a key. Add an extra box off each of the consequences to **explain** the impact it would have. Start with 'This means that...'

Case study: Flooding in Boscastle, Cornwall, UK

Background

The Cornish village of Boscastle was hit by a **flash flood** on 16 August 2004. Two billion litres of water travelled at speeds of 65 kilometres per hour through the village. Seven centimetres of rain fell in only two hours, making it a 1 in 400-year event.

Causes

A number of physical and human factors contributed to the flash flood.

Physical factors:
- Torrential rain (over 60 mm in two hours) and a rising high tide caused river levels to increase by 2.15 metres in one hour.
- The unusually heavy rainfall was linked to the remains of Hurricane Alex, which had travelled across the Atlantic Ocean to the UK.
- High land in the area causes warm, moist air coming onshore to rise rapidly. As the air cooled and condensed it formed dense cumulonimbus clouds up to 12,000 metres high.

- The ground was already saturated from above-average rainfall in the previous weeks, causing water to flow straight into the river channel.
- Boscastle is at the confluence of three rivers. High volumes of water converged on the one village.
- Steep hillsides caused water to funnel quickly into the narrow river valley.
- The river basin is very small at only 23 square kilometres.

Human factors:
- The linear layout of the historic fishing village, with houses situated either side of the narrow river channel, did not help.
- Cars and vegetation were trapped under low-lying bridges and acted as a dam, forcing water to find an alternative route.
- The volume of water overwhelmed old sewer systems.

Consequences

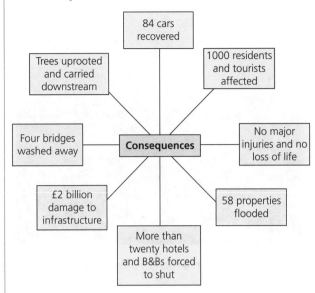

Figure 15 The aftermath of the flood

Responses

The responses can be divided into short-term and long-term responses.

Figure 16 Responses to the Boscastle flood

Short term	Long term
Use of helicopters to rescue people from rooftops	River Jordan culvert to divert water into the lower part of the river, nearer to the sea
Within a few days, diggers had cleared the streets of debris and uprooted trees	Widening and deepening the river channel
	Removing low bridges and raising the height of new bridges
Use of sandbags to keep as much water as possible out of shops and houses	Raising the height of the car park
	Retaining and strengthening revetments
	Widening the river further upstream to slow the flow and encourage large sediment to deposit
Hotels and B&Bs welcomed the tourists whose cars were swept out to sea	Removing trees adjacent to the river and reinstating meadows to store more water and slow the flow into rivers

Now test yourself

TESTED ☐

1 Explain how the Boscastle flood was caused by a combination of human and physical factors.
2 Sort the consequences of the flood shown in Figure 15 into social, economic and environmental.
3 Which of the responses do you think are the most and least sustainable?

Flash flood: a sudden flood, typically due to heavy rainfall.

Revision activity

This case study has a lot of facts and figures. Create your own pairs game. Cut up 20 small pieces of card and put a statistic on one, then what that number means on another. Lay them all face down and turn two at a time. When you have a pair, move them to one side. Why not time yourself to see how quickly you can match the pairs?

Case study: Heatwave in the UK

Background

In late June and early July 2015, there was a short **heatwave** in the UK. Temperatures were hotter than in Rome and Athens. On 1 July 2015, temperatures hit 36.7 °C in Heathrow, the highest July temperature on record.

Causes

Light winds across the UK helped to draw in hot air from the high-pressure system that was over central and southern Europe, where a heatwave had already been declared.

Consequences

The heatwave only lasted a few days, so the impacts were minimal. However:

● Wimbledon fans were advised to protect themselves from the high temperatures with hats and sunscreen.
● Some schools cancelled their sports day.
● Train tracks began to buckle and road surfaces started to melt.
● Car breakdown call-outs were up fourteen per cent.
● Barbecue sales went up 67 per cent, and sunglasses by 29 per cent.

Figure 17 Brighton beach in July 2015

Responses

Many of these responses can also be considered to be **secondary impacts** as they are a direct result of the **primary impacts** listed above:

● Network Rail set speed restrictions on lines that were vulnerable to buckling in the heat.
● Trains were cancelled.
● 999 calls doubled in one day, particularly from the elderly.
● The government issued a Level 3 Heatwave Action alert.

Revision activity

Copy the list of consequences and highlight them with three different colours to indicate which are social, economic and environmental. Add a key.

Now test yourself

TESTED ☐

1 Describe how heatwaves can bring both positive and negative consequences.
2 Describe how a high-pressure system can lead to a heatwave (use information from page 9 to help you with this question).

Heatwave: a prolonged period of abnormally hot weather.

Primary impacts: occur as a direct result of the hazard.

Secondary impacts: occur as a result of the primary impacts.

Case study: Typhoon Haiyan in the Philippines

Background

Super typhoon Haiyan hit the south-east coast of the Philippines, an emerging and developing country (EDC), with winds of up to 312 kilometres per hour on 8 November 2013. The regions of Leyte, Tacloban and Samar were the worst affected.

The Philippines is made up of 7000 islands and experiences around twenty typhoons a year. It is a multiple hazard zone, which means that it suffers from a range of other hazards, such as volcanic eruptions, landslides and flooding.

Responses

The responses can be divided into short-term and long-term responses.

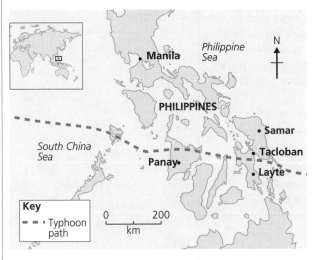

Figure 18 Path of Typhoon Haiyan

Causes

This region has ideal conditions for tropical storms to form:
- a vast expanse of warm waters across the Pacific Ocean
- low wind shear, keeping the structure of a storm intact
- the large number of small, dispersed islands are not big enough to reduce the energy of the typhoon.

Typhoon Haiyan caused a five-metre storm surge with waves up to fifteen metres high. It was so large in diameter that it covered two-thirds of the country.

> **Super typhoon**: a storm that reaches sustained wind speeds of at least 240 kilometres per hour.

Human factors

The low level of development in the Philippines affects the country's ability to cope with natural disasters of this scale. Much of the country lives in extreme poverty and there has been rapid population growth, particularly in vulnerable coastal areas. In Tacloban, the population grew from 76,000 to 221,000 in 40 years. Buildings and storm shelters are poorly constructed.

Consequences

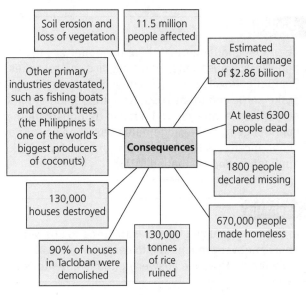

Figure 19 Consequences of Typhoon Haiyan

Responses

Different groups responded to Typhoon Haiyan:
- The charity ShelterBox provided people with water purification equipment, blankets and solar-powered lights.
- 1215 evacuation centres were set up.
- Britain sent HMS *Daring* as part of the emergency response.
- The Philippine government deployed soldiers to restore law and order.
- The UK government contributed £50 million in aid.
- The Philippine Red Cross delivered basic food aid including rice and canned goods.

The relief effort was hampered by heavy rain and damage to roads and infrastructure, which had made some remote areas inaccessible. It was difficult to assess the damage and distribute aid effectively due to the number of small islands.

Now test yourself

TESTED ☐

1 How did the geography of the Philippines contribute to the disaster?
2 How are the consequences linked to the economic development of the Philippines (an EDC)?
3 Which were greater, the social or economic consequences? Justify your answer.

Revision activity

Copy the spider diagram in Figure 19. Use three different colours to indicate the social, economic and environmental consequences. Add a key.

Exam practice

Evaluate the significance of the causes of a UK-based weather hazard event. [6]

ONLINE ☐

Exam tip

A very common mistake is to use an incorrect case study. Check and double-check this before you start to write. Exam practice question 1 requires a UK-based case study. Some students will make the mistake of writing about a non-UK based example, such as Typhoon Haiyan.

Exam tip

The most challenging parts of the paper are the extended 6-mark and 8-mark questions, which require you to 'assess' or 'evaluate'. Evaluation involves judgement and opinion. You will be expected to make some comment about how important, significant or valuable something is, and you will need to demonstrate that you have the confidence to make judgements based on your knowledge. The mark scheme reflects this by awarding marks for both knowledge (AO1) and application and evaluation (AO3).

Exam tip

In the exam you will have 6–8 minutes to answer a 6-mark question. This is not very much time to carry out an evaluation of your knowledge. Practise answering these questions using a timer.

Exam tip

Annotate the question so that you know exactly what you are being asked to do. Looking at Exam practice question 1, for example:

Make judgements on ...explanation

What made it happen? Human/physical factors?

Evaluate the **significance** of the **causes** of a **UK-based** weather hazard event

What was the *main* cause? Which other factors had a part to play?

2004 Boscastle flood or 2015 heatwave

If you do not annotate the question you are likely to make the mistake of writing a lengthy answer that includes everything you know about the case study, as opposed to writing a focused response that answers the specific question given.

2 How do plate tectonics shape our world?

Plate tectonics

The structure of the Earth

REVISED

The Earth's internal structure is divided into three layers: the **core**, **mantle** and **crust** (Figure 1). The crust and upper mantle are called the lithosphere. The lithosphere is broken into seven large sections and twelve smaller sections called **tectonic plates**. Tectonic plates are rigid and move very slowly, floating across the heavier molten rock in the mantle.

There are two types of plate:
- **continental plates**, which are thicker (25–100 km) but less dense
- **oceanic plates**, which are thinner (5–10 km) but much denser.

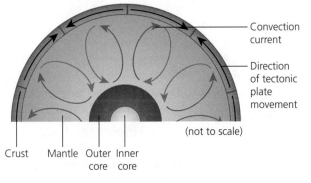

- Convection current
- Direction of tectonic plate movement

(not to scale)

Crust Mantle Outer core Inner core

Figure 1 The Earth's structure

> **Core**: the centre of the Earth; it has a solid metal inner core (at a temperature of 6000 °C) and semi-solid outer core (4030–5730 °C) from which heat is radiated outwards through the mantle.
>
> **Mantle**: hot, dense liquid rock (magma); it moves continuously due to heat from the core (convection), which drives plate movement.
>
> **Crust**: the solid, rocky shell layer (lithosphere) over the mantle around the Earth, on which the continents and oceans sit; the Earth's crust is fragmented into tectonic plates that float on the mantle.

How do tectonic plates move?

REVISED

1. The hot core (6000 °C) heats the magma. This heat makes the magma less dense so it rises in the mantle towards the crust.
2. As the magma rises it cools towards the crust. The magma becomes denser and sinks back towards the core.
3. This causes **convection currents** in the mantle.
4. Convection currents build pressure and cause the tectonic plates, which are floating on the mantle, to be pushed and pulled along in different directions.
5. A **plate boundary** is where two or more plates meet (Figure 2). Some plates collide (**destructive** or **collision** plate boundaries), some separate from each other (**constructive** plate boundaries) and others move past each other (**conservative** plate boundaries).

> **Tectonic plates**: the crust is broken up into seven large sections and various smaller sections, which are floating on the mantle; they move towards, away from and past each other.
>
> **Continental plates**: the lithosphere (crust) that is underneath our continents and land.
>
> **Oceanic plates**: the lithosphere (crust) that is underneath our oceans.

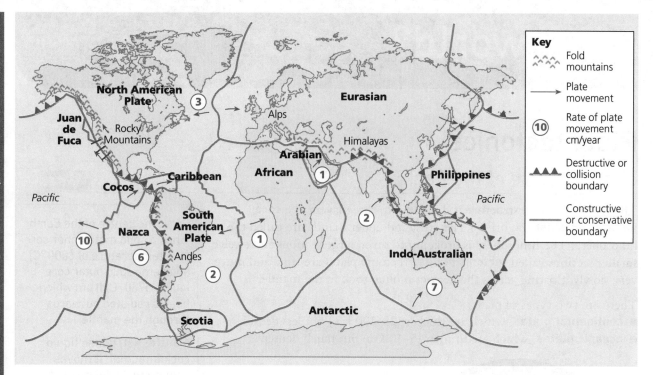

Figure 2 World map of plate boundaries

Convection currents: the constant churning of the mantle due to heat energy (radiation) passing out from the core.

Plate boundary: where two or more tectonic plates meet; mountain building and hazards such as earthquakes and volcanoes can be found at these boundaries.

Destructive plate boundary: where two plates are converging or coming together and the oceanic plate is subducted; it can be associated with violent earthquakes and explosive volcanoes.

Collision plate boundary: where two plates are converging or coming together and continental plates are pushed upwards to form mountains.

Constructive plate boundary: where rising magma adds new material to plates that are diverging or moving apart.

Conservative plate boundary: where two tectonic plates slide past each other.

Now test yourself

TESTED

1 Why do convection currents occur in the mantle?
2 Using the map in Figure 2:
  (a) Identify where there are conservative boundaries.
  (b) What type of plate margin is found between the UK and the USA?

Plate boundaries

Processes at different types of plate boundaries

There are four types of plate boundary (Figure 3, pages 25 and 26). Each is associated with different landforms and hazards.

The distribution of **earthquakes** and **volcanoes** is not random: the majority occur in narrow bands along plate boundaries. Earthquakes are found at all types of plate boundaries. Volcanoes are found at constructive and destructive plate boundaries. There are anomalies (data values that do not match the pattern of a sample), however, as some occur in the middle of plates at **hotspots** (see page 26).

> **Earthquake**: a sudden or violent movement within the Earth's crust, followed by a series of aftershocks.
>
> **Volcano**: an opening in the Earth's crust from which lava, ash and gases erupt.

Figure 3 Characteristics of different plate boundaries

Plate boundary	Direction of plate movement	Physical process	Hazards
Constructive	Plates move away from each other or diverge **Example:** the Eurasian and North American plates	As tectonic plates move away from each other, pressure is released, allowing hot, molten magma to rise to the surface. Magma erupts through fissures and faults, causing volcanoes.	**Earthquakes:** Yes (usually small, not violent) **Volcanic eruptions:** Yes (shield)
Destructive	Plates move towards each other or converge **Example:** the Pacific and Philippine plates	When tectonic plates converge, pressure builds between them. The rock eventually fractures, causing earthquakes. When oceanic and continental plates collide, the denser oceanic plate subducts under the continental plate into the mantle, where it melts. Hot magma can rise through the lithosphere and erupt as lava through volcanoes.	**Earthquakes:** Yes (violent) **Volcanic eruptions:** Yes (composite)
Collision	Plates move towards each other or converge **Example:** the Indian and Eurasian plates	When tectonic plates converge, pressure builds between them. The rock eventually fractures, causing earthquakes. When two continental plates collide, the rock folds and crumples up, forming fold mountains such as the Himalayas. There is no subduction.	**Earthquakes:** Yes **Volcanic eruptions:** No

Figure 3 Characteristics of different plate boundaries

Plate boundary	Direction of plate movement	Physical process	Hazards
Conservative 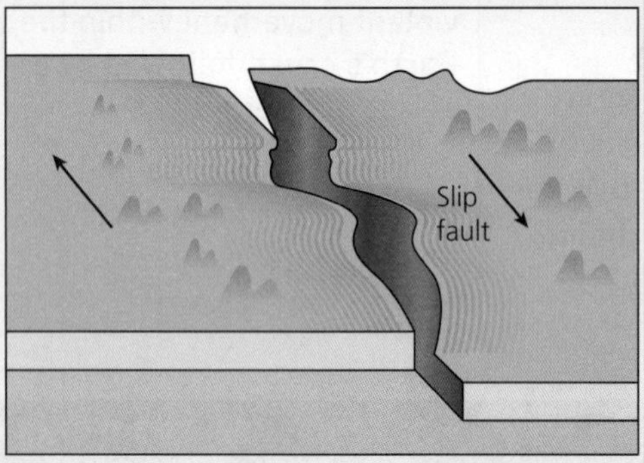	Plates slide parallel past each other **Example:** the Pacific and North American plates	Plates slide past each other, either in opposite directions or in the same direction but at different speeds. Friction builds between the plates until the friction is finally overcome and the rock fractures and jolts forwards, causing an earthquake.	**Earthquakes:** Yes **Volcanic eruptions:** No
Hotspots 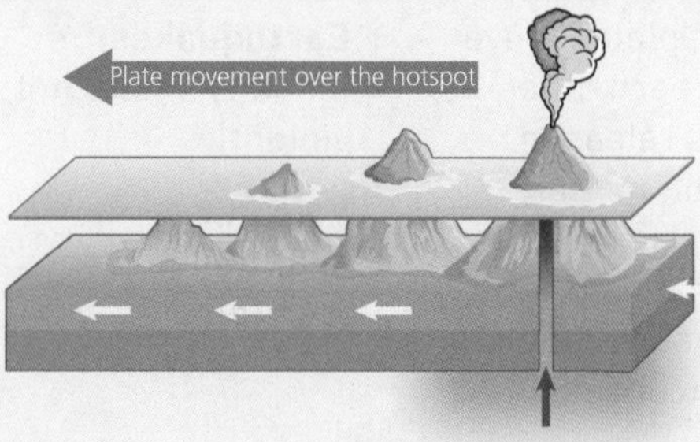	Not to do with plate boundaries **Example:** Hawaii and the Azores	When a tectonic plate moves over a particularly hot area of the mantle, a super-heated plume of hot magma rises up towards the crust. It breaks through the plate where it is thin enough or at a fracture. Magma erupts to the surface. Volcanic islands can appear.	**Earthquakes:** Yes **Volcanic eruptions:** Yes

Hotspots: weaknesses in the Earth's crust where the rock is thinner; this allows magma to the surface even though it is not at a plate boundary.

Now test yourself

TESTED ☐

1 Which plate boundaries experience:
   (a) earthquakes (b) volcanoes?
2 How are the plates moving at the following plate boundaries:
   (a) constructive (b) destructive
   (c) collision (d) conservative?
3 Where can volcanoes be found other than at plate boundaries?

Revision activity

Draw four simple sketches of the different plate boundaries. Annotate them with the information from Figure 3.

Make a set of sort cards to practice matching up the definitions on this page and their meanings.

Exam practice

1 What do collision and destructive plate boundaries have in common? [1]
2 Outline the differences between conservative and destructive plate boundaries. [2]
3 Explain how volcanoes occur at destructive plate boundaries. [4]

ONLINE ☐

Earthquakes and volcanoes

How does plate movement cause earthquakes?

An earthquake is a violent shaking of the Earth's crust. It is caused by a sequence of steps:

1 Convection currents within the Earth's mantle cause tectonic plates to move.
2 Pressure builds between the plates due to friction.
3 Eventually the pressure that has built up reaches its limit.
4 The plates suddenly jolt as the rock breaks at the **focus**.
5 The energy is released through seismic waves (waves of energy passed through the Earth or along its surface due to plate movement) that travel out through the rock. This is an earthquake.

Focus: the location in the Earth where an earthquake starts.

Epicentre: the point on the Earth's surface vertically above the focus.

Advanced countries: countries that share a number of important economic development characteristics, including well-developed financial markets, high degrees of financial intermediation and diversified economic structures with rapidly growing service sectors.

Depth of earthquakes

The point where the rock first breaks in the Earth's crust is the focus. The **epicentre**, which lies directly above the focus, is the point on the Earth's surface where the earthquake is first felt. The focus can occur at different depths within a tectonic plate.

● A shallow focus occurs closer to the Earth's surface (0–70 km). It is likely to cause more surface damage.
● A deep focus occurs further away from the Earth's surface (70–700 km). It typically causes less damage as the seismic waves have further to travel.

As shown in Figure 4, the closer to the epicentre of an earthquake, the greater the effects are likely to be, as the energy released by the earthquake will be at its strongest. Damage caused by earthquakes may be affected by other factors such as:

● **geology (rock type):** softer rock such as clay and sand will shake more easily
● **building design and infrastructure:** the quality of materials and foundations can have an impact on the extent and nature of damage
● **level of economic development:** less developed countries often have lower economic losses but a higher death toll; this is usually reversed in **advanced countries**
● **time of day:** an earthquake at rush hour in an urban area can have very severe effects
● **population density and distance from the epicentre:** if more people live close to the epicentre, where seismic waves are stronger, it is likely that more people will be affected by the earthquake.

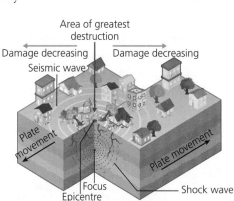

Figure 4 Conservative plate movement causes earthquakes

Measuring earthquakes

The energy released by an earthquake is measured using a seismometer on the moment magnitude scale or the Richter scale. The Mercalli scale measures the intensity of the impact caused by the earthquake. It is based on the perception of the effects on humans, buildings and the environment.

Exam practice

1 Draw an annotated diagram to explain why earthquakes occur at conservative plate boundaries. [4]
2 Distinguish between a shallow and a deep focus. [2]

How does plate movement cause volcanoes?

A volcano is an opening in the Earth's crust through which lava, ash, steam, rock particles and gas erupt from the mantle. The type of volcano depends on whether it occurs at a destructive or a constructive plate boundary, or at a hotspot.

Shield volcanoes

Convection currents cause tectonic plates to move away from each other at constructive plate boundaries. This allows magma to rise from the mantle and erupt to the surface. The lava is runny. This forms a **shield volcano**, which is lower with gentle slopes (Figure 5).

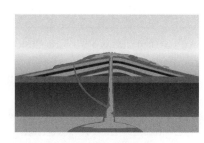

Figure 5 Shield volcano

Composite volcanoes

At destructive plate boundaries convection currents cause tectonic plates to move towards each other. The denser oceanic plate is forced down into the subduction zone in the mantle. The plate melts and the magma rises to form a **composite volcano** (Figure 6). The lava is more viscous and sticky. This forms a volcano that is taller with steep sides. It is built up in layers every time an eruption occurs. Composite volcanoes are also called stratovolcanoes.

Figure 6 Composite volcano

Hotspot volcanoes

Hotspots occur when an oceanic plate is moving over a particularly hot area of the mantle, which creates a super-heated plume of hot magma rising up towards the crust. This can break through the plate where it is thin enough or where there are fractures in the rock. This leads to the appearance of volcanic islands. As the oceanic plate keeps moving away from the hotspot, the material will stop being fed to the volcano and it will become extinct, and a new volcano will form over the hotspot. The further from the hotspot an island is, the older it is (Figure 7).

> **Shield volcano**: a gentle-sloped volcano with runny lava; they are found at constructive plate boundaries.
>
> **Composite volcano**: a steep-sided volcano made up of layers; they are found at destructive plate boundaries.

Revision activity

Sketch diagrams of a shield and composite volcano. Label their characteristics.

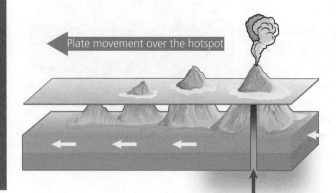

Plate movement over the hotspot

Figure 7 Hotspot volcanoes appear as islands

Exam practice

1 Describe the differences in the shape of shield and composite volcanoes. [3]
2 Describe how hotspot volcanoes form. [4]
3 Explain how plate movement causes a composite volcano to form. [6]

ONLINE

Now test yourself

1 What type of plate boundary does a shield volcano form on?
2 Describe the lava at a composite volcano.
3 Why do hotspot volcanoes form in a chain?

TESTED

Case study: A tectonic hazard event

Case study: Eyjafjallajökull in Iceland

Causes

Eyjafjallajökull is located in Iceland. It lies on a constructive plate boundary of the North American and Eurasian plates.

- **19 March 2010:** increasing seismic activity was monitored. Earthquakes became more frequent and shallower as magma rose. The magma chamber filled up inside the volcano.
- **20 March 2010:** lava erupted through fissures. Ash began to be deposited on the surface of the glacier as lava broke through the ice. Magma deeper in the mantle continued rising and mixing, causing chemical reactions. An increase in silica and gases meant that the magma was more viscous and explosive.
- **12 April 2010:** a second fissure ripped open and lava began flowing again.
- **15 April 2010:** lava burst through a 200-m fissure of thick glacial ice, causing flooding. Lava and water mixed creating a huge plume of tephra rising 10,000 m into the atmosphere.

Consequences

- Twenty farms were destroyed.
- Increased local revenue generated by stranded tourists.
- Respiratory health problems for animals and humans.
- The international economy suffered as the ash cloud led to the cancellation of 95,000 flights worldwide. Passengers and cargo were stranded. Airlines lost $200 million a day. Stock market shares in air travel agencies dropped four per cent. Kenya lost $3.5 million in cancelled trade and perishable food decaying. Europe lost $2.8 billion in insurance costs and lost trade.
- Petrol prices in the UK and Europe increased as the oil industry lost income with 1.87 million barrels of unused air fuel.
- Increased use of Eurostar, train services, ships and ferries.
- Glaciers were covered in dark ash for months, which increased melting.
- Melting glacial ice destroyed parts of Iceland's Route 1.

Responses

- Local farm residents and livestock were evacuated.
- The Iceland Meteorological Office sent text messages, radio, television and internet alerts to the public.
- Iceland's well-trained national emergency agency replaced bridges with pre-built temporary structures, dredged blocked rivers and cleared many tonnes of ash.
- Aircraft were grounded according to weather predictions.

Now test yourself

TESTED ☐

1 Which plate boundary is Eyjafjallajökull found on?
2 What does 'hazardous' mean?
3 What were the indicators that an eruption was imminent?
4 Identify two local and two international consequences.
5 Why was evacuation necessary?

Exam practice

1 Explain how the responses to the 2010 Eyjafjallajökull eruption reduced the impact of the hazard. [6]
2 'Volcanic eruptions are always negative in their consequences.' To what extent do you agree with this statement? [9]

ONLINE ☐

Revision activity

Draw a table with the following three headings: Economic, Social and Environmental. List the consequences of the 2010 eruption of Eyjafjallajökull under these headings.

Exam tip

Use facts and figures that you have learnt in case study questions to prove that you are writing specifically about that place and not any other.

Impact of technology in hazard zones

Eight per cent of the world's population live near a volcano. Fifty per cent of the USA's population live in earthquake-prone areas.

Technology can aid **monitoring**, **prediction**, **protection** and **planning** strategies to **mitigate** the risks associated with living in hazard zones.

Monitoring: recording physical changes, such as earthquake tremors around a volcano, to help forecast when and where a natural hazard might strike.

Prediction: attempts to forecast when and where a natural hazard will strike based on current knowledge; this can be done to some extent for volcanic eruptions (and tropical storms) but less reliably for earthquakes.

Protection: actions taken before a hazard strikes to reduce its impact, such as educating people or improving building design.

Planning: actions taken to enable communities to respond to, and recover from, natural disasters; measures include emergency evacuation plans, information management, communications and warning systems.

Mitigation: the action of trying to reduce the impact of a hazard by planning, predicting and preparation (for example, building earthquake-resistant buildings).

Prediction
REVISED

Predicting the time, date and exact location of an earthquake is extremely difficult as there is little warning. Seismometers (instruments used to detect and record earthquakes) and computer modelling are used to monitor earthquake-prone areas. There has been no successful prediction yet, however.

Volcanic eruptions are easier to predict as there are advance warning signals. Nevertheless, the exact time and day are unknown.

Media, especially social media, is used to distribute information immediately and automatically to inform people of the risks and advise on how best to respond, for example turning off utilities or evacuations.

Early warning systems
REVISED

Earthquakes:
- Radon gas is monitored as it may be released before rock fractures.
- Seismometers and GPS are used to monitor foreshocks.
- Electromagnetic waves can be passed through rock to try to detect developing fractures.

Volcanic eruptions:
- Tiltmeters and GPS are used to measure deformation of ground.
- Seismometers measure earthquake patterns near the magma chamber.
- Satellite imagery and thermal scanning shows rising magma.
- Thermal heat sensors record ground and river temperatures.
- Radon and sulphur gas levels are monitored using gas-trapping bottles.

Building design
REVISED

- Earthquake-proof buildings can be designed, although they are expensive (e.g. rubber shock absorbers, automatic shutters).
- Sea walls offer some protection against possible tsunamis.

- Buildings can be insulated and barriers put in place to reduce ash entry. Roofs can be reinforced to cope with the weight of ash without collapsing.

Now test yourself
TESTED

1 What makes predicting an earthquake difficult?
2 How is media used to save lives in hazard zones?
3 Name three indicators that a volcano is soon to erupt.

Exam practice

1 Describe how technology used in building design can reduce the damage caused by an earthquake. [4]
2 Explain how technology is used to predict a volcanic eruption. [6]

ONLINE

Now test yourself and exam practice answers at **www.hoddereducation.co.uk/myrevisionnotes**

3 What evidence is there to suggest climate change is a natural process?

The pattern of climate change

Climate change during the Quaternary period

REVISED

The Earth is believed to be 4.55 billion years old. The period of time that stretches from 2.6 million years ago to the present day is called the **Quaternary period**. The entire Quaternary period is often called an **ice age** due to the presence of a permanent ice sheet on Antarctica.

There has been **climate change** during the Quaternary period. Temperatures have fluctuated wildly but overall have gradually cooled. There have been cold 'spikes', which are known as glacial episodes. In between each cold spike are warmer inter-glacial episodes. Today we live in an inter-glacial episode. The average temperature today is higher than almost all of the Quaternary period, as can be seen in Figure 1.

> **Quaternary period**: the most recent geological period covering the last 2.6 million years, during which time there were several cold and warm periods.
>
> **Ice age**: a glacial episode characterised by lower than average global temperatures and during which ice covers more of the Earth's surface.
>
> **Climate change**: changes in long-term temperature and precipitation patterns that can either be natural or linked to human activities.

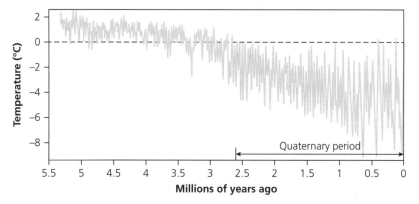

Figure 1 Average global temperatures for the last 5.5 million years

Climate change during the last 400,000 years

REVISED

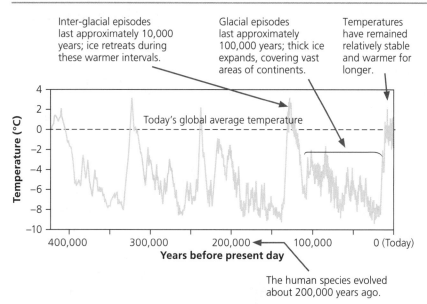

Inter-glacial episodes last approximately 10,000 years; ice retreats during these warmer intervals.

Glacial episodes last approximately 100,000 years; thick ice expands, covering vast areas of continents.

Temperatures have remained relatively stable and warmer for longer.

Today's global average temperature

The human species evolved about 200,000 years ago.

Figure 2 Trends in average global temperatures (400,000 years ago to the present day)

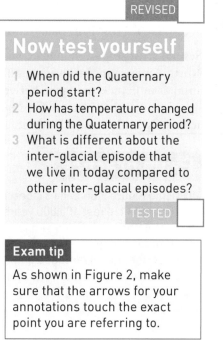

Now test yourself

1 When did the Quaternary period start?
2 How has temperature changed during the Quaternary period?
3 What is different about the inter-glacial episode that we live in today compared to other inter-glacial episodes?

TESTED

> **Exam tip**
>
> As shown in Figure 2, make sure that the arrows for your annotations touch the exact point you are referring to.

Evidence for climate change

Since 1914, the Met Office has recorded reliable climate change data using weather stations, satellites, weather balloons, radar and ocean buoys. They have collected the following evidence:
- an increase in average surface air temperature by 1 °C over the last 100 years
- the warmest ocean temperatures since 1850 have been measured
- an average rise in sea levels of 20 cm since 1900.

Evidence for past climate change

REVISED

Evidence for past climate change can be gathered from a range of sources:
- **Sea ice positions**: maps and photos show that many of the world's glaciers are retreating as temperatures rise and ice melts. Arctic sea ice has declined in volume by ten per cent in the last 30 years. The Muir Glacier (Alaska, USA) has retreated 50 kilometres in the last 120 years. Data recording is considered very reliable; it only goes back a relatively short period of time, however.
- **Ice cores**: oxygen, carbon dioxide and methane in ice cores can help estimate past temperature (spanning 800,000 years) by comparing it to present levels. Scientists drill deep into the ice in the Antarctic and Greenland to extract ice that is thousands of years old. The data is considered very reliable.
- **Global temperature data**: NASA uses over 1000 ground weather stations and satellite information to map global temperature. Average global temperatures have increased by 0.6 °C since 1950 and 0.85 °C since 1880. However, weather stations are not evenly distributed – there are fewer especially in Africa – so reliability could be questioned. Computer programs are used to produce global temperature maps, which does not necessarily make them reliable. In addition, data only goes back to 1880.
- **Paintings and diaries**: diaries and written observations can suggest evidence of climate change at the time, such as:
  - price increases in grain in Europe
  - sea ice preventing ships from landing in Iceland
  - people emigrating due to crop failures
  - winter 'Frost Fairs' held on the frozen River Thames.

Several artists captured much colder winter landscapes in Europe and North America in the seventeenth century than today. Cave paintings of animals in France and Spain between 11,000 and 40,000 years ago show significant climate change. However, it is difficult to date cave paintings accurately, and personal accounts and art are subjective viewpoints.

Now test yourself

1 List the different ways in which evidence about climate change is collected.
2 How do paintings show evidence of climate change?
3 Give one problem with using the following as evidence for climate change:
   (a) sea ice position
   (b) global temperature data.

TESTED

Exam practice

1 Using Figure 2 on page 31, describe the pattern of temperature change in the last 400,000 years. [4]
2 Explain how changing sea ice positions provide evidence of climate change. [4]
3 Describe the advantages and disadvantages of using paintings and diaries as evidence of climate change. [4]

ONLINE

Exam tip

Circle any plurals in exam questions. This will help you to notice when you need to consider more than one factor. Full marks cannot be gained unless you have obeyed this in questions.

Causes of natural climate change

There is evidence that climate change occurred before humans existed. This means that climate change must be a natural phenomenon. However, natural causes alone cannot account for the unprecedented temperature increase since the 1970s.

Sunspots and volcanic eruptions

Sunspots are darker patches on the Sun's surface. They are caused by magnetic activity inside the Sun. Sunspots increase from a minimum number to a maximum number in a sunspot cycle of about every eleven years. Scientists suggest that the more sunspots there are, the more heat is given off by the Sun. However, solar output from the Sun has barely changed in the last 50 years so it cannot be responsible for the climate change seen since the 1970s.

> **Sunspot:** a spot or dark patch that appears from time to time on the surface of the Sun; they are associated with an outburst of energy from the Sun.

Volcanic eruptions throw huge quantities of ash, gases (including sulphur dioxide) and liquids into the atmosphere. When sulphur dioxide mixes with water vapour it becomes a volcanic aerosol. Volcanic aerosols reflect sunlight away, and this reduces global temperatures. Wind carries material far beyond where it was ejected from the volcano, so the reduced temperatures are also experienced elsewhere.

Milankovitch cycles

The distribution of the Sun's energy on the Earth varies due to changes in the Earth's orbit. The cyclical time periods that relate to the Earth's orbital changes around the Sun are called **Milankovitch cycles**. There are three of them: axial title, precession and eccentricity.

> **Milankovitch cycles:** the cyclical time periods that relate to the Earth's orbital changes around the Sun.

Axial tilt

The Earth spins on its tilted axis. The angle of the tilt changes due to the gravitational pull of the Moon. When the angle of the tilt is greater, it is associated with a higher average temperature. The angle of the tilt moves back and forth every 41,000 years.

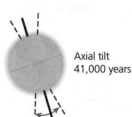

Axial tilt
41,000 years

Figure 3 Axial tilt

Precession

The Earth is not a perfect sphere; as the Earth spins it wobbles on its axis in a 26,000-year cycle.

Eccentricity

The Earth's orbit around the Sun is not fixed and changes over time from being almost circular to being mildly elliptical. The cycle takes 100,000 years. Colder periods occur when the Earth's orbit is more circular, and warmer periods occur when it is more elliptical.

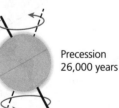

Precession
26,000 years

Figure 4 Precession

Now test yourself

1 Identify two natural factors causing climate change.
2 What is in a volcanic eruption that reduces global temperatures?
3 Why can volcanic eruptions cause lower temperatures in other regions away from the volcano?
4 What takes:
   (a) 11 years
   (b) 26,000 years
   (c) 41,000 years
   (d) 100,000 years?

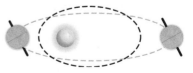

Figure 5 Eccentricity

The natural greenhouse effect

What is the greenhouse effect?

The **natural greenhouse effect** is a naturally occurring phenomenon that keeps the Earth warm enough for life to exist. Without it, the Earth would be about 33 °C colder. This is the process:

1 The Sun's infrared heat rays enter the Earth's atmosphere.
2 The heat is reflected from the Earth's surface.
3 The natural layer of greenhouse gasses allows some heat to be reflected out of the Earth's atmosphere but some is trapped. This keeps the Earth warm enough.

> **Natural greenhouse effect:** the natural warming of the planet as some of the heat reflected by the Earth is absorbed by liquids and gases in the atmosphere, such as carbon dioxide.

The enhanced greenhouse effect

Scientists have proved that natural causes are responsible for climate change, yet they cannot account for the increases in temperature since the 1970s (Figure 8). Humans must also be contributing to the greenhouse effect.

Human activity has increased the layer of greenhouse gases that exits naturally. The thicker layer of greenhouse gases (77 per cent carbon dioxide, 14 per cent methane, 8 per cent nitrous oxides, 1 per cent Chlorofluorocarbons (CFCs)) means that less of the Sun's energy is able to escape the Earth's atmosphere, so the temperature increases even more. This is the **enhanced greenhouse effect**.

> **Enhanced greenhouse effect:** the exaggerated warming of the atmosphere caused by human activities, resulting in the natural greenhouse effect becoming more effective.

> **Revision activity**
>
> Draw a pie chart to show the contribution of each of the greenhouse gases.

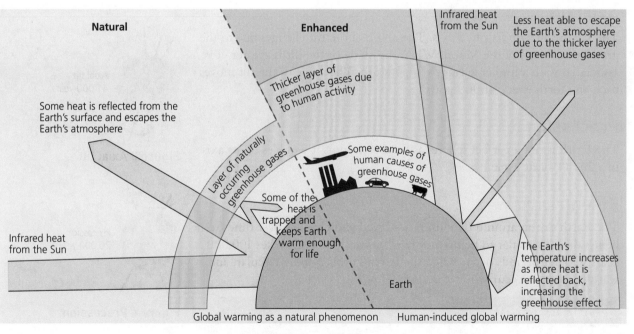

Figure 6 The greenhouse effect: natural and enhanced

Humans enhance the greenhouse effect by generating more greenhouse gases through activities such as those shown in Figure 7.

Figure 7 How human activities contribute to the main greenhouse gases

Greenhouse gas	Human activities that contribute to the enhanced greenhouse effect
Carbon dioxide (CO_2)	Burning fossil fuels (coal, oil and gas)
	Deforestation (burning wood)
	Industrial processes (for example, making cement)
Methane (CH_4)	Emitted from livestock and rice cultivation
	Decay of organic waste in landfill sites
Nitrous oxides (NO_x)	Vehicle exhausts
	Agriculture and industrial processes

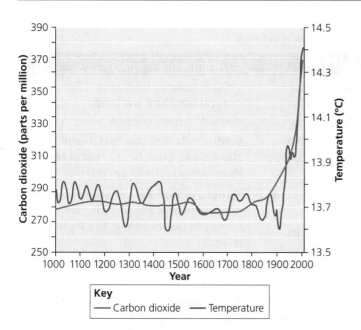

Figure 8 Carbon dioxide and global temperature change

Now test yourself

TESTED

1 Name the greenhouse gases.
2 Why does the Earth need the natural greenhouse effect?
3 Identify two human activities thought to be causing climate change.
4 Name the fossil fuels.

Exam practice

1 Using Figure 8, describe the relationship between carbon dioxide and temperature. [2]
2 Explain the enhanced greenhouse effect. [4]
3 Explain why natural factors cannot be solely responsible for climate change. [6]

ONLINE

Impacts of climate change worldwide

The effects of climate change are not certain. They are likely to be unevenly distributed across the world and will depend on the human and physical circumstances of the location. For example, low-lying coastal countries will be more vulnerable to effects such as flooding, and poorer countries will be more vulnerable as they have less ability to invest in prediction and protection strategies.

Sea level rise and extreme weather events

REVISED

- **Sea level rise:** The Intergovernmental Panel on Climate Change (IPCC) reported that the sea level has risen by an average of 20 cm since 1900. It could rise by another 26–82 cm by 2100.

- **Extreme weather events:** Single extreme weather events cannot be linked to long-term climate change. However, scientists suggest the increasing frequency of extreme weather events can be blamed on climate change.

Sea level rise

Figure 9 Social, economic and environmental impacts of sea level rise

Social impacts	Economic impacts	Environmental impacts
600 million live in coastal areas less than 10 metres above sea level	**Valuable agricultural land** lost to the sea or polluted by seawater (for example, Bangladesh, Vietnam)	IPCC estimates that up to 33% of **coastal land and wetlands** could be lost in the next 100 years
Increase in **environmental refugees** due to flooding (for example Tuvalu and Vanuatu)	Many **world cities**, including New York and London, could be affected by flooding	Biodiversity lost due to damage by storms and bleaching in **coral reefs** (for example, the Great Barrier Reef)
Job losses in fishing or tourism, so have to **learn new skills**	**Transport infrastructure** damaged by flood water	**Mangrove forests**, which form natural barriers to storms, are damaged (for example, the Pacific Islands)
Migration and **overcrowding** in low-risk areas due to flooding (for example in Asia)	**Investment in coastal defences** required as UK's current defences are under increased pressure from sea level rise	**Fresh water sources polluted** by salty seawater
	Loss of income from **tourism** as beaches eroded or flooded; hotels forced to shut	**Adélie penguins** on the Antarctic Peninsula may decline as ice retreats

Extreme weather events

Figure 10 Social, economic and environmental impacts of extreme weather events

Social impacts	Economic impacts	Environmental impacts
Increased **drought**, affecting farming and water supplies (for example, California 2012–17)	Extreme weather increases investment in **prediction and protection**	**Forests** experience more pests, disease and forest fires (for example South East Australia had worst bushfires on record in 2009)
Increased risk of **diseases** such as skin cancers and heatstroke as temperatures increase	Flood risk increases **repair and insurance costs** (for example, damage of $9.7 billion in Pakistan in 2010)	Lower rainfall causes **food shortages** for orang-utans in Borneo and Indonesia
Winter-related **deaths** decrease with milder winters	Maize **crop yields** decrease by up to 12% in South America; they will increase in northern Europe but require more irrigation	**Flooding** in South Asia increases rice yields
	Skiing industry may decline in Alps as less snow	

Now test yourself and exam practice answers at **www.hoddereducation.co.uk/myrevisionnotes**

Impacts of climate change within the UK

Impact on weather patterns

Summers are expected to become drier, but winters will receive an increase in rainfall. Some rivers will flood more frequently in winter. Temperatures are set to increase, but increases are expected to be greater in the south of the UK.

Figure 11 Impact of climate change on weather patterns in the UK

Economic	Environmental	Social
The summer heat will lead to growth in the tourist industry in the Lake District, generating jobs and increased revenue. The Cairngorms ski resorts may be forced to close. This may reduce revenue.	Vegetation and ecosystems will move north. Sitka spruce yield may increase in Scotland and new crops such as peaches and oranges could be cultivated in southern England. Agricultural productivity may increase under warmer conditions but would require increased irrigation.	The UK's elderly will be more vulnerable during heatwaves but will suffer fewer cold-related deaths in winter. Heating costs will reduce. Water shortages would be experienced by many by the 2050s.

Impact on seasonal patterns

Spring is expected to arrive earlier and autumn start later. Precipitation is expected to become even more seasonal.

Figure 12 Impact of climate change on seasonal patterns in the UK

Environmental
Bird migration patterns will shift. Some trees and plants will flower earlier and others later. Wildlife species could struggle to survive if the seasons do not match up with their food supply.

Impact on sea level

The UK is expected to be at a greater risk of coastal flooding due to sea level rise.

Figure 13 Impact of climate change on sea level in the UK

Economic	Environmental	Social
The Thames Barrier will need expensive upgrading or need to be replaced due to the increased risk of flooding in the Thames Estuary. Teesside industries on coastal mudflats will be vulnerable to sea level rise. Agricultural land may be lost due to managed retreat. The tourism industry could be affected by eroded beaches.	Salt marshes may become flooded and eroded; however, managed retreat could create new salt-marsh habitats.	Cliff collapse may increase, putting properties at risk.

Now test yourself

1 How is rainfall and temperature changing in the UK as a result of climate change?
2 How are seasons in the UK changing as a result of climate change?
3 Is the risk of coastal flooding expected to increase or decrease as a result of climate change?
4 State three examples of changes in industry in the UK as a result of climate change.
5 List two social and two environmental impacts of climate change in the UK.

Exam practice

1 Identify a negative and positive environmental impact of climate change in the world. [2]
2 Describe the social and economic impacts of climate change in the UK and around the world. [6]
3 Explain how the environment is affected by climate change in the UK. [6]

3 What evidence is there to suggest climate change is a natural process?

4 What makes a landscape distinctive?

The concept of a landscape

How landscapes are defined

Landscapes are the visible features that make up the surface of the Earth. They exist at a variety of scales, from local (such as a valley or city park) to regional (such as the Pennines) and even international (such as the Alps or Andes mountains).

There are four elements that make a landscape (Figure 1):
- **Natural/physical:** the geology of an area and the physical processes that are operating both now and in the past (such as rivers and glaciers).
- **Biological:** the soils and vegetation that have colonised the area together with the natural ecosystems (plants, animals, fungi, and so on).
- **Human:** people have had a huge impact on the landscape, constructing roads, urban areas, service infrastructure (such as pylons, wind and solar farms) and altering land use (for example deforestation, afforestation, farming and golf courses).
- **Variables:** temporary features of a landscape such as weather events, clouds, changing colours and smells.

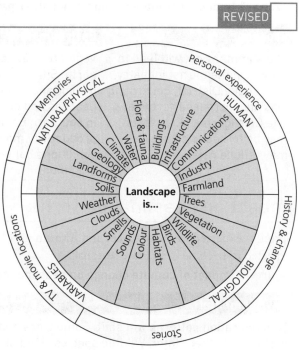

Figure 1 What is a landscape?

It is possible to identify both natural and built landscapes.
- **Natural landscapes**: an example could be a water feature such as a river or a stream. Natural landscapes are predominantly devoid of human activity, such as roads, railways, industry and other urban developments. Of course, there are very few truly natural landscapes in the UK as most of the countryside has been managed by people (through farming and planting woodland).
- **Built landscapes**: these are essentially towns or cities, comprising houses, shopping centres, industries and transport networks. Some people consider the expansion of the built environment as a threat to natural environments (such as wind farms, new transport infrastructure and modern housing developments). This can lead to conflict between different groups of people.

> **Natural landscapes**: a landscape that is the result of natural processes and has not been shaped or changed by human activity.
>
> **Built landscapes**: the human-made surroundings that provide the environment for human activity; it may also refer to towns, cities and other urban environments.

Now test yourself

TESTED

1 Define the term 'landscape'.
2 Explain the difference between a natural landscape and a built landscape.
3 Explain why there are few truly natural landscapes in the UK.

Revision activity

Draw up a table to identify the different elements (natural/physical, biological, human and variable) of a landscape that you know well.

The distribution of landscapes in the UK

How are landscapes distributed in the UK?

The UK enjoys a huge variety of distinctive physical landscapes, including spectacular mountains (such as the Lake District in northern England and the Cairngorms in Scotland), rolling hills and valleys (much of southern England) and flat wetlands (such as the Somerset Levels).

Figure 2 is an atlas map showing the distribution of uplands and lowlands in the UK. Notice that most of the mountainous uplands in northern England, Wales and Scotland were glaciated in the past when the climate was very much colder than it is today. Immense glaciers carved spectacular valleys and created the dramatic landscapes that we associate with these mountainous areas.

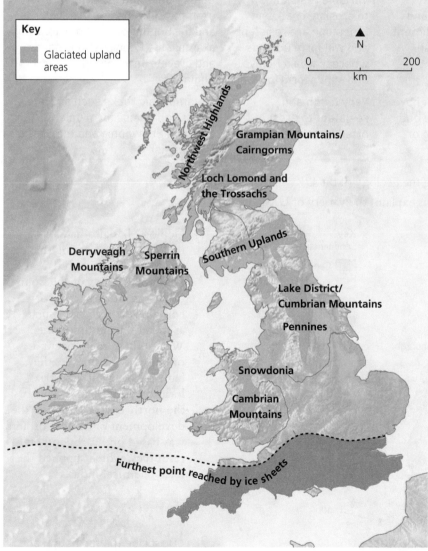

Figure 2 The distribution of upland, lowland and glaciated areas in the UK

The characteristics of landscapes in the UK

Geology

Geology is all about the rocks beneath our feet. It is one of the most important factors affecting the physical and human landscapes of the UK. It is possible to identify three types of rock: igneous, sedimentary and metamorphic.

Figure 3 Formation, characteristics and examples of the three rock types

Rock type	Formation	Characteristics	Examples
Igneous	Magma that has cooled either on the ground surface (extrusive) when a volcano erupts or below the ground (intrusive)	Tough and resistant to erosion; igneous rocks often form uplands	Dartmoor (granite), Northern Ireland (basalt), Cuillin Hills (gabbro)
Sedimentary	Rocks formed from the accumulation and compaction of sediment, usually in the ocean	Variable resistance to erosion; chalk and limestone are resistant and will form uplands whereas weaker clays and sands form lowlands	Chalk ridges and escarpments (for example Lincolnshire, the Chilterns and the South Downs), limestone (for example the Pennines), and sands and gravels (lowlands in southern England)
Metamorphic	Existing rocks that have undergone change due to extreme heating or pressure	Very tough and resistant to erosion, often forming uplands	Metamorphic rocks such as slate, schist and gneiss form uplands (for example Snowdonia and the Scottish mountains)

Figure 4 is a simplified map showing the geology of the UK. The great range of colours representing different rocks explains the variety of UK landscapes.

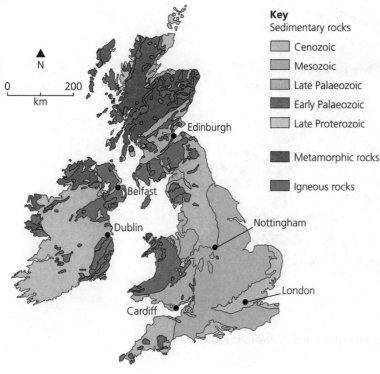

Key

Sedimentary rocks

- Cenozoic
- Mesozoic
- Late Palaeozoic
- Early Palaeozoic
- Late Proterozoic

- Metamorphic rocks

- Igneous rocks

Geology has also had an influence determining the location of built landscapes. Some rocks are valuable sources of energy (such as coal) or contain raw materials (such as metal ores in limestone). Where these rocks occur at or close to the surface, they have encouraged industrial location and urbanisation. In the north-east of England, the development of Middlesbrough was based on nearby mineral resources, which supplied the local chemical industry.

Figure 4 Simplified geological map of the UK

Now test yourself

Describe the formation and landscape characteristics of igneous, sedimentary and metamorphic rocks.

Climate

Climate is the long-term average weather conditions, calculated over a period of 30 years. It differs from the weather, which is the day-to-day condition of the atmosphere (temperature, humidity, wind speed, and so on).

The climate of the UK has a role to play in creating the UK's distinctive landscapes, but it is also affected *by* the landscapes – it is a two-way relationship.

The UK enjoys a maritime climate, with the prevailing (most common) winds blowing across the Atlantic from the south-west. This accounts for the generally high rainfall and moderate temperatures throughout the year.

- **Rainfall**: the uplands – particularly in the west – receive a high proportion of the rainfall, with the lowlands in the south and east tending to be drier. This is because the moist air from the Atlantic is forced to rise and cool over the western uplands, forming rain-bearing clouds. This type of rainfall is called relief rainfall. The drier regions to the east can be described as being in the 'rain shadow'.
- **Temperatures**: temperatures tend to be lower in the uplands than in the lowlands, with frost and snow being common hazards in the winter. This is because temperature falls on average by 0.6 °C per 100 metres of altitude.

Climate affects physical processes such as weathering and erosion.

- The process of freeze–thaw is very active in upland areas, resulting in jagged rock surfaces and accumulations of scree on mountainsides.
- Rivers are fast flowing (due to high rainfall) and very erosive in uplands, carving deep V-shaped valleys.
- In the past, extreme cold in the uplands led to the formation of ice caps and glaciers, which carved spectacular landscapes.

Human activity

Human activity has transformed the landscape of the UK from a largely forested landscape at the end of the last ice age (about 10,000 years ago) to the present day's agricultural and urbanised landscape.

- **Uplands**: these areas are sparsely populated due to the harsh climate and steep relief. Human activity is limited to extensive sheep rearing and forestry. Reservoirs have been created in some areas to supply water and to generate hydroelectric power. In recent years, wind farms have been constructed in some upland areas, exploiting the strong winds.
- **Lowlands**: these areas are more densely populated due to the moderate climate and gentler relief. Commercial farming dominates the countryside and much of this landscape is urbanised or criss-crossed by transport and service infrastructure.

> **Revision activity**
>
> Construct a spider diagram to show how geology, climate and human activity affect UK landscapes (uplands and lowlands).

Exam practice

1 Outline the main characteristics of the built environment? [2]
2 Describe how human activities have affected the landscape of the UK. [4]
3 Explain how geology and climate have affected the upland landscapes of the UK. [6]

ONLINE

5 What influences the landscapes of the UK?

Geomorphic processes

Geomorphic processes are responsible for shaping landscapes. They include weathering, mass movement, erosion, transportation and deposition.

Weathering

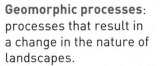

Weathering involves the decomposition or disintegration of rock in its original place at or close to the ground surface. There are two main types of weathering: **chemical weathering** and **mechanical (physical) weathering**.

Figure 1 Processes of chemical and mechanical weathering

Chemical weathering	Mechanical weathering
Carbonation: carbon dioxide dissolved in rainwater forms a weak carbonic acid; this reacts with calcium carbonate (limestone and chalk) to form calcium bicarbonate, which is soluble and can be carried away by water	**Freeze–thaw**: repeated cycles of freezing and thawing, causing water trapped in rocks to expand and contract, eventually causing rock fragments to break away (Figure 2)
Hydrolysis: acidic rainwater reacts with feldspar in granite, turning it into clay and causing granite to crumble	**Salt weathering**: crystals of salt, often evaporated from seawater, grow in cracks and holes, expanding to cause rock fragments to flake away
Oxidation: oxygen dissolved in water reacts with iron-rich minerals, causing rocks to crumble	

A third type, **biological weathering**, involves living organisms such as nesting birds, burrowing rabbits and plant roots. Plant roots may expand in cracks, slowly prising rocks apart. Acids that promote chemical weathering may be active beneath soils and rotting vegetation.

Now test yourself

TESTED

1 What is the difference between mechanical and chemical weathering?
2 Outline the process of freeze–thaw weathering.
3 How does the action of plant roots cause weathering to rocks?

☀ Day

Water collects in cracks in the rock

☾ Night

Water freezes to form ice
Expansion causes stresses and the cracks are enlarged

☀/☾ Repeated freezing and thawing

Rock fragment breaks off and collects as scree

Figure 2 Freeze–thaw weathering

Mass movement

Mass movement is active at the coast, particularly where cliffs are undercut by the sea, making them unstable. It includes sliding and slumping, as well as falls (rockfalls) and flows (mudflows).

- **Sliding**: this involves rock or loose material sliding downhill along a slip plane, such as a bedding plane. Slides are often triggered by ground shaking (for example, an earthquake) or heavy rain.
- **Slumping**: this commonly involves the collapse of weak rock, such as sands and clays, often found at the coast. Slumping often results from heavy rainfall when the sediments become saturated and heavy.

Common forms of mass movement at the coast include:

- **Rockfall:** individual fragments or chunks of rock falling off a cliff face, often resulting from freeze–thaw weathering.
- **Landslide:** blocks of rock sliding rapidly downslope along a linear shear-plane, usually lubricated by water (Figure 3).
- **Mudflow:** saturated material (usually clay) flowing downhill, which may involve elements of sliding or slumping as well as flow.
- **Rotational slip/slump:** slumping of loose material often along a curved shear-plane lubricated by water (Figure 4).

> **Mass movement**: movement of surface material caused by gravity.

Figure 3 Landslide

> **Revision activity**
>
> Make a copy of Figure 3 and add labels to describe the causes and characteristics of sliding.

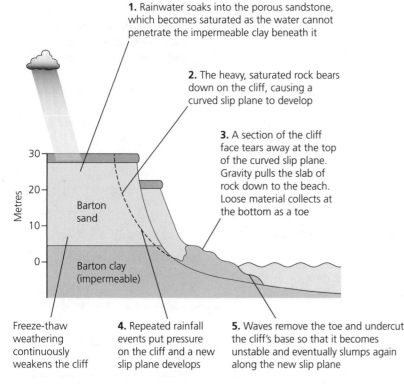

1. Rainwater soaks into the porous sandstone, which becomes saturated as the water cannot penetrate the impermeable clay beneath it

2. The heavy, saturated rock bears down on the cliff, causing a curved slip plane to develop

3. A section of the cliff face tears away at the top of the curved slip plane. Gravity pulls the slab of rock down to the beach. Loose material collects at the bottom as a toe

Barton sand

Barton clay (impermeable)

Freeze-thaw weathering continuously weakens the cliff

4. Repeated rainfall events put pressure on the cliff and a new slip plane develops

5. Waves remove the toe and undercut the cliff's base so that it becomes unstable and eventually slumps again along the new slip plane

Figure 4 Slumping at Barton on Sea, Hampshire

Now test yourself

Use Figure 4 to describe the causes and characteristics of slumping.

Erosion

REVISED

Erosion involves the wearing away and removal of material by a moving force, such as a breaking wave. The main processes of erosion are **abrasion**, **hydraulic action**, **attrition** and **solution**.

> **Erosion**: the wearing away and removal of material by a moving force.

Abrasion: this is the 'sandpapering' effect as loose rock particles carried by the water scrape against solid bedrock. It can also involve loose particles being flung against a sea cliff or river bank by the water – a process sometimes referred to as corrasion.

Hydraulic action: this involves the sheer power of the water, often compressing air into cracks in sea cliffs or river banks, causing rocks to break away.

Attrition: erosion caused when rocks and boulders transported by waves bump into each other and break up into smaller pieces. Over time the rocks become smaller and more rounded.

Solution: the dissolving of soluble rocks, such as chalk and limestone.

Transportation

REVISED

Transportation involves the movement of eroded sediment from one place to another. It commonly involves the following processes: **traction**, **saltation**, **suspension** and **solution**.

> **Transportation**: the movement of eroded sediment from one place to another.

Traction: large particles rolling along the seabed.

Saltation: a bouncing or hopping motion by pebbles too heavy to be suspended.

Suspension: particles suspended within the water.

Solution: chemicals dissolved in the water.

> **Revision activity**
>
> Draw a simple diagram to show the four processes of transportation. Set your diagram in a river or at the coast.

Deposition

REVISED

Deposition occurs when material being transported is dropped due to a reduction in energy. This typically occurs in areas of low energy, where velocity is reduced and sediment can no longer be transported. At the coast, deposition is common in bays or in areas sheltered by bars and spits. In rivers, deposition is common close to the river banks, in estuaries and at the inside bend of meanders.

> **Deposition**: when material being transported is dropped due to a reduction in energy.

Exam practice

1 Describe the process of mechanical weathering. [2]
2 Explain the conditions under which hydraulic action will be an important process of coastal erosion. [4]
3 Explain where and why sediment is deposited in rivers and at the coast. [6]

ONLINE

Coastal landforms

Headlands and bays

REVISED

Headlands and **bays** are characteristic features of a discordant coastline where rocks of different hardness (resistance to erosion) are exposed at the coast. Headlands form resistant promontories jutting out into the sea. They are separated by bays of less resistant rock where the land has been eroded back by the sea.

Look at Figure 5. Notice that coastal erosion processes erode away the weaker rocks more readily than the harder rocks to form a sequence of alternating headlands and bays.

> **Headland:** an area of land that extends out into the sea, usually higher than the surrounding land.
>
> **Bay:** an area of the coast where the land curves inwards.

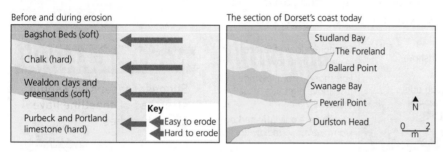

Figure 5 Formation of headlands and bays on the Dorset coast

Erosion at the headlands creates landforms such as cliffs and wave-cut platforms. Deposition occurs in the sheltered waters of the bay, forming a beach.

Caves, arches and stacks

REVISED

Caves, **arches** and **stacks** are commonly formed at headlands where relatively tough rock juts out into the sea. They are formed in several steps:
1 Cracks in the cliff (joints or faults) are eroded and enlarged by hydraulic action to form a wave-cut notch at the base of the cliff. This is eroded further by hydraulic action and abrasion to form a cave.
2 The cave is eroded right through the headland to form an arch.
3 Over time, processes of erosion widen the arch and processes of weathering weaken its roof.
4 Eventually the roof collapses to form an isolated pillar of rock called a stack. Over time the stack will be eroded and will collapse to form a stump, which is only exposed at low tide.

> **Cave:** a natural underground chamber or series of passages, especially with an opening to the surface; also refers to the extended cracks at the base of a cliff.
>
> **Arch:** an arch-shaped structure formed as a result of natural processes within a rock feature such as a cliff.
>
> **Stack:** a coastal feature that results from erosion; a section of headland that has become separated from the mainland and stands as a pillar of rock.

Revision activity

Draw a series of annotated diagrams to show the formation of a stack.

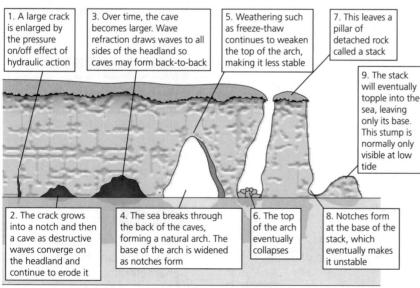

Figure 6 Formation of caves, arches and stacks

Now test yourself

Explain the formation of an arch.

TESTED

Beaches

A **beach** is a depositional landform made of sand or pebbles (shingle) extending from the low water line to the upper limit of storm waves. Beaches may exhibit a range of small landforms such as ridges called berms (formed by waves just above the high tide line, for example), ripples or shallow water-filled depressions called runnels. Figure 7 shows a typical beach profile.

- **Sandy beaches**: commonly formed in sheltered bays, associated with relatively low-energy constructive waves. Flat and extensive beaches are often backed by sand dunes.
- **Pebble beaches**: commonly associated with higher-energy coastlines where destructive waves remove finer sand, leaving behind coarser pebbles. Beaches tend to be steep and narrow with distinctive high tide berms.

Beach: a depositional landform made of sand or pebbles extending from the low water line to the upper limit of storm waves.

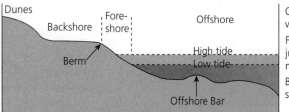

Offshore – lowest level of wave action to low tide

Foreshore – from low tide to just above high tide, usually marked by a berm

Backshore – only affected by storm waves so mostly dry

Figure 7 Beach profile

Suggest reasons why some stretches of coastline have sandy beaches.

Spits

A **spit** is a sand or shingle (pebble) ridge most commonly formed by **longshore drift** operating along a stretch of coastline (Figure 8). It is rather like a narrow beach extending out from the coast into the sea. At its end, where it is more exposed to variations in wind and waves, it tends to curve to form a hook or recurved tip. Some spits have sand dunes on them.

Spit: a sand or shingle ridge most commonly formed by longshore drift operating along a stretch of coastline.

Longshore drift: the movement of sediments along a stretch of coastline as a result of wave action.

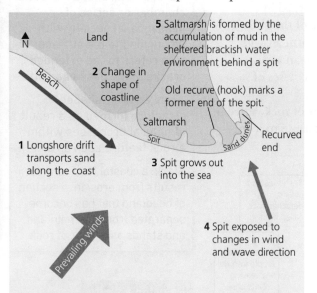

1 Longshore drift transports sand along the coast

2 Change in shape of coastline

3 Spit grows out into the sea

4 Spit exposed to changes in wind and wave direction

5 Saltmarsh is formed by the accumulation of mud in the sheltered brackish water environment behind a spit

Old recurve (hook) marks a former end of the spit.

Saltmarsh

Spit

Recurved end

Sand dunes

Land

Beach

Prevailing winds

N

Figure 8 Formation of a spit

Remember that a spit is land, so it lies above the high tide line. On an Ordnance Survey map, use the high tide line to trace the edges of a spit.

Make up your own version of Figure 8 with longshore drift operating from north to south along a stretch of coastline. Ensure that your diagram is fully labelled.

1 Describe how rock type can lead to the formation of headlands and bays. [4]
2 Explain the formation of caves, arches and stacks. [6]
3 Explain how longshore drift leads to the formation of a spit. [6]

Make use of annotated diagrams when describing the characteristics or formation of coastal landforms.

River landforms

V-shaped valley

A **V-shaped valley**, as the term implies, is a steep-sided, narrow river valley that takes the form of a V-shape in its cross profile. It is formed by river erosion as the river cuts down vertically into the mountainous landscape. Weathering and mass movement on the valley sides are responsible for broadening the top of the valley profile. Interlocking spurs – 'fingers' of land jutting into the valley – are common landforms in the upper course of a river (Figure 9).

> **V-shaped valley**: a steep-sided, narrow river valley that takes the form of a V-shape in its cross profile.

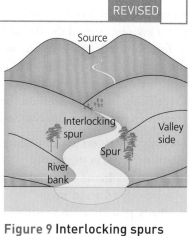

Figure 9 Interlocking spurs

Waterfalls

A **waterfall** is typically formed where fast-flowing water plummets over a vertical cliff – often a considerable drop – into a deep plunge pool below (Figure 10). Hydraulic action and abrasion are mainly responsible for eroding the plunge pool, which can be several metres deep in the centre.

Waterfalls most commonly form when a river flows over a hard, resistant band of rock. Unable to erode the tougher rock, a 'step' is formed in the long profile of the river.

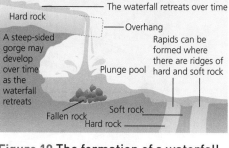

Figure 10 The formation of a waterfall

> **Waterfall**: a steep fall of river water where its course crosses between different rock types, resulting in different rates of erosion.

Gorges

A **gorge** is commonly formed by the upstream retreat of a waterfall (Figure 11).

- Erosion (primarily hydraulic action and abrasion) undercuts the hard rock forming the waterfall to create an overhang.
- Eventually this overhang collapses into the plunge pool, causing the waterfall to retreat upstream.
- Over many thousands of years, this process of undercutting and collapse continues and a gorge is formed immediately downstream.

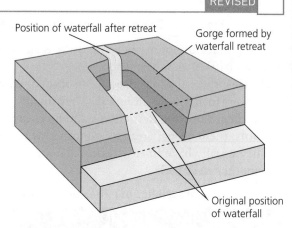

Figure 11 The formation of a gorge

> **Gorge**: a narrow valley between hills or mountains.

Meanders

Meanders are commonly found in a river's middle and lower course, where they can form extensive and elaborate bends. The fastest-flowing water swings around the outside bend of a meander, eroding the banks to form a river cliff. Here the water is deep. On the inside bend, where the velocity is lower, deposition occurs, forming a slip-off slope. In this way, the meander develops an asymmetrical cross profile (Figure 12). Over time, lateral erosion on the outside bend widens the river valley and creates an extensive, flat **floodplain**.

Meandering rivers are most commonly associated with the following environmental conditions:
- gentle gradients
- relatively fine sediments
- steady precipitation regime throughout the year.

This explains why they are by far the most common river pattern in the UK and, indeed, throughout the world.

> **Meander**: a bend in a river that results from the flow of water along it.
>
> **Floodplain**: the flat area of land either side of a river channel forming the valley floor, which may be flooded.

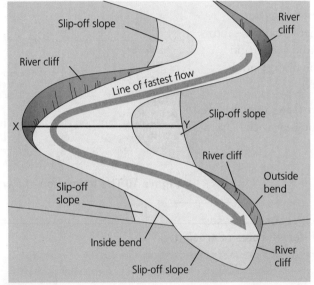

Cross profile

Figure 12 Characteristics of a river meander

A common feature of meandering rivers is an alternating pattern of shallows (riffles) and deeps (pools).
- **Riffles**: these shallow areas are associated with the straighter sections of rivers in-between meanders. They usually have rocky beds and turbulent flow due to friction with the river bed.
- **Pools**: these deeper areas are associated with the meander bends. They usually have finer sediment and less turbulent flow due to the smoother river bed.

> **Now test yourself**
>
> Draw a sketch cross profile from X to Y on Figure 12 and explain the formation of the features of erosion and deposition.
>
> TESTED

Figure 13 Characteristics of riffles and pools

Rate of flow	Riffle	Pool
High flow (winter)	Greater friction in the shallower riffles results in slower, more turbulent flow	Water tends to flow faster through the deeper pools due to a reduction in friction with the bed and banks
Low flow (summer)	On entering a riffle, the reduction in channel size often results in slightly faster flow	The reduced volume of water tends to slow down on entering a deep pool, where the channel is larger

Oxbow lakes

Oxbow lakes are typically seen with rivers in their middle or lower course. They represent old meander bends that have been cut off by faster flowing water during times of flood (Figure 14). Gradually, without water flowing through them, they will become filled with silt and marshy vegetation will grow.

> **Oxbow lake**: a horseshoe-shaped lake that forms when a meander is separated from the main river channel as a result of erosion.

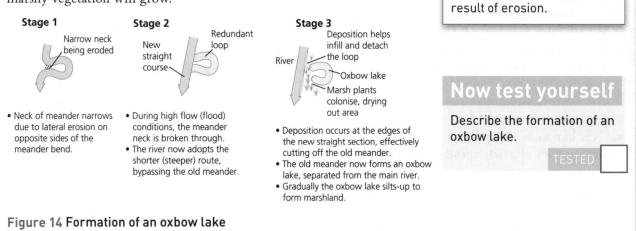

Stage 1
Narrow neck being eroded

- Neck of meander narrows due to lateral erosion on opposite sides of the meander bend.

Stage 2
New straight course
Redundant loop

- During high flow (flood) conditions, the meander neck is broken through.
- The river now adopts the shorter (steeper) route, bypassing the old meander.

Stage 3
Deposition helps infill and detach the loop
River
Oxbow lake
Marsh plants colonise, drying out area

- Deposition occurs at the edges of the new straight section, effectively cutting off the old meander.
- The old meander now forms an oxbow lake, separated from the main river.
- Gradually the oxbow lake silts-up to form marshland.

Now test yourself

Describe the formation of an oxbow lake.

TESTED

Figure 14 Formation of an oxbow lake

Levees

Levees are raised river banks commonly found in the lower course of rivers. They are formed during flood conditions when water flows over the river banks on to the surrounding floodplain (Figure 15) in a sequence of steps:

> **Levee**: raised banks along a river.

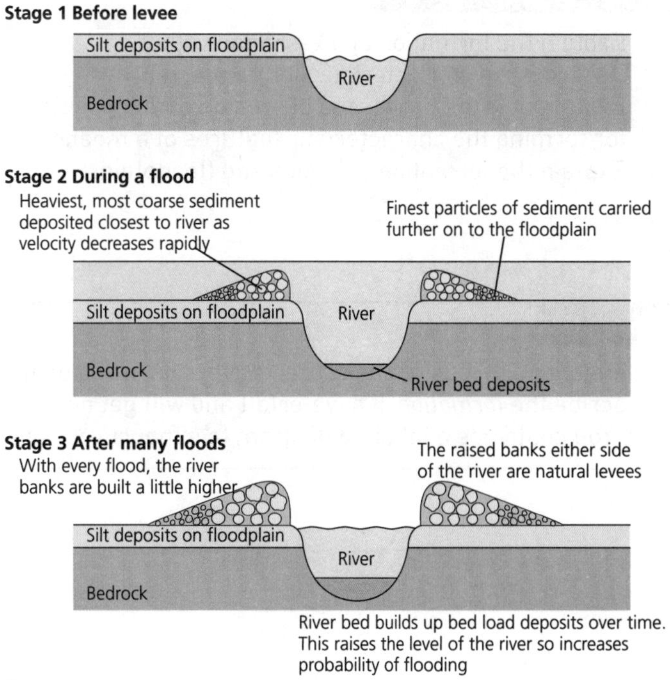

1 As water overtops the river banks, there is a sudden localised reduction in the velocity of the water, which had previously been flowing very fast along the river channel.
2 This causes sediment in suspension to be deposited at the river bank.
3 Coarse (heavier) sediment is deposited first and this traps the finer sediment.
4 With each successive flood, the deposited sediment raises the river banks by as much as a few metres.

Levees can be created artificially by people to contain water within a river channel to reduce the threat of flooding. In the USA, the term 'levee' is most commonly used for the artificial form.

Stage 1 Before levee
Silt deposits on floodplain
River
Bedrock

Stage 2 During a flood
Heaviest, most coarse sediment deposited closest to river as velocity decreases rapidly
Finest particles of sediment carried further on to the floodplain
Silt deposits on floodplain
River
Bedrock
River bed deposits

Stage 3 After many floods
With every flood, the river banks are built a little higher
The raised banks either side of the river are natural levees
Silt deposits on floodplain
River
Bedrock
River bed builds up bed load deposits over time. This raises the level of the river so increases probability of flooding

Figure 15 Formation of a levee

Now test yourself

TESTED

Use a series of simple diagrams to explain the formation of a levee.

Floodplains

Floodplains are associated with rivers in their middle and lower course. They are extensive, flat areas of land mostly covered by grass. There may be some marshy areas close to the river, and also oxbow lakes (Figure 16). As the name implies, they are formed during flood conditions and are periodically and quite naturally inundated by water.

- During a flood, water containing large quantities of alluvium (river silt) pours out over the flat valley floor.
- The water slowly soaks away, leaving the deposited sediment behind.
- Over hundreds of years, repeated flooding forms a thick alluvial deposit that is fertile and often used for farming.

Floodplains become wider due to the lateral erosion of meanders.

- When the outside bend of a meander meets the edge of the river valley, erosion will cut into it, thereby widening the valley at this point.
- As meanders slowly migrate downstream, the entire length of the valley will eventually be widened.

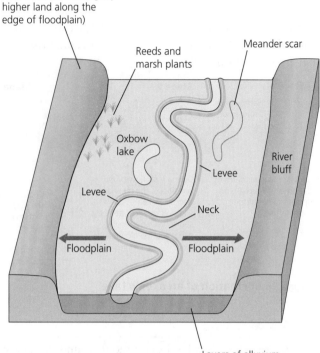

Figure 16 Characteristic features of a floodplain

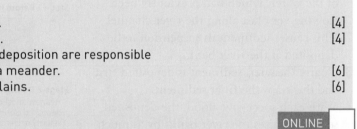

Exam practice

1 Explain the formation of a V-shaped valley. [4]
2 Describe the characteristics of a waterfall. [4]
3 Explain how the processes of erosion and deposition are responsible for forming the characteristic features of a meander. [6]
4 Explain the formation of levees and floodplains. [6]

Exam tip

In question 2, focus on the *characteristics* only. You do not need to describe the *formation* of a waterfall and will get no marks for doing so. You could use a labelled diagram to support your answer.

Case study: A coastal landscape

In this section you need to study **one** coastal landscape. You should focus on the following aspects:

● geomorphic processes operating at different scales; how they are influenced by geology and climate
● landforms created by geomorphic processes
● the impact of human activity (including management) on geomorphic processes and the landscape.

Case study: North Norfolk coast

The North Norfolk coast is located in East Anglia in the east of England (Figure 17). It is a low-lying stretch of coastline with a variety of habitats including broad sandy beaches, salt marshes, sand dunes and some stretches of cliffed coastline, particularly east of Cromer. The landscape is famous for its 'big skies', attracting tourists and artists.

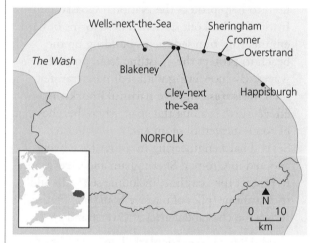

Figure 17 The North Norfolk coast

Geomorphic processes

Geomorphic processes operate at different scales, both spatially (space) and temporally (time). The following list includes a few of the geomorphic processes seen on the North Norfolk coast:

● The North Norfolk coast is exposed to powerful waves from the north-east, the direction of maximum fetch (over 4000 km). When the winds come from the north-east, waves are at their most powerful.
● Longshore drift predominantly operates from east to west along this stretch of coastline, although opposing currents transport sediment southwards along the east coast of England, around the Wash.
● Mechanical weathering processes, such as freeze–thaw and mass movement (particularly slumping), are very active on the cliffs in the east of the region, particularly between Overstrand

and Happisburgh. Unlike longshore drift, these processes are more localised and smaller scale.
● Coastal erosion is active in the east, with coastal deposition dominating in the west.

How are geomorphic processes affected by geology and climate?

Geology: The entire region is underlain by the sedimentary rock chalk, which is exposed in places such as the base of the cliffs at Overstrand. Overlying the chalk is thick glacial sediment deposited by ice sheets that spread south from Scandinavia during the last glacial period.

● Cromer Ridge – a 100 m ridge just inland from Cromer – is a terminal moraine, a glacial deposition that marks the furthest extent of ice advance.
● Sands and gravels deposited by glacial meltwater streams are found along parts of the coast to the west of the region.
● Till – unsorted glacial sediment – forms thick deposits along much of the coast, spreading some distance inland.

These glacial sediments are extremely weak. They are prone to mass movement and are rapidly eroded by the sea. In parts of the east of the region, cliffs are retreating by over 1 m each year.

Climate: In geological terms, changes in climate have had a significant impact on the North Norfolk coast. During the last glacial period, ice advanced over the area, depositing huge thicknesses of sediment. Exposed to geomorphic processes, this has been rapidly eroded in the east and redistributed in the west.

The present-day climate of the region is relatively dry, with warm summers and occasionally cold winters. Till dries out and cracks in dry conditions, making it more vulnerable to geomorphic processes. Cold spells in the winter will promote freeze–thaw, particularly if the clay contains deep cracks.

Coastal landforms

The North Norfolk coast exhibits landforms of coastal erosion and coastal deposition. The north-east coast between Overstrand and Happisburgh is dominated by actively eroding till cliffs. In Happisburgh, some people have lost their homes due to rapid coastal retreat. Figure 18 shows the typical profile of these cliffs.

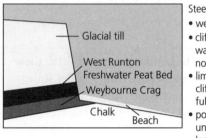

Glacial till

West Runton
Freshwater Peat Bed

Weybourne Crag

Chalk

Beach

Steep cliffs due to:
- weak rock (till)
- cliffs exposed to powerful waves and long fetch to north-east
- limited beach, leaving cliffs exposed to the full force of waves
- powerful waves undercutting the cliff leading to collapse.

Figure 18 Typical North Norfolk cliffs

The north coast to the west of Sheringham is a coastline dominated by deposition. Longshore drift transports sediment from east to west, forming an extensive spit at Blakeney (Blakeney Point). Sand dunes (for example, at Holkham) and salt marshes (for example, at Stiffkey) are also features of this stretch of coastline.

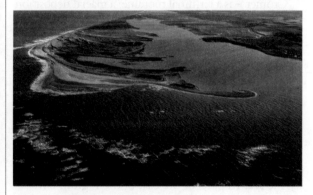

Figure 19 Blakeney Point Spit

Impact of human activity and management

The North Norfolk coast is widely used by people:
- There are many villages and small towns on the coast, linked by main roads. Economic activities include fishing, farming and forestry.
- Much of the area is popular with tourists who visit the coast to enjoy the beautiful landscapes, taking boat trips and hiking or cycling along the lanes and paths.
- The varied and undamaged coastal habitats, such as the salt marshes and sand dunes, are particularly popular with visitors.

Like other coastal regions in England and Wales, the North Norfolk coast is managed to balance the social, economic and environmental demands on the coast. Planners face several physical challenges, including high rates of coastal erosion, occasional storm surges (most recently in 2013) and long-term sea level change associated with climate change. A number of coastal defence schemes have been introduced to address the issues:

- At Holkham the local landowner (the Holkham Estate) has planted pine trees to help stabilise the sand dunes. The Estate has also constructed boardwalks to enable visitors to access the dunes without damaging the vegetation or disturbing the wildlife.
- At Wells-next-the-Sea, groynes (barriers constructed perpendicular to the coast) have been constructed to protect the beach huts; gabions (metal cages filled with rocks) have been used to help protect the National Coastwatch Institution lookout station. These forms of hard engineering trap sediment, building up the beach to protect the coastline from powerful waves. By interfering with sediment transfer, such measures can have harmful knock-on effects further along the coast where beaches can become starved of sediment.
- Several hard engineering measures have been adopted in Cromer, Sheringham and Overstrand to protect the coastline, including sea walls and rock armour (piles of massive boulders). Figure 20 shows the coastal defences at Overstrand.

Coastal defences – particularly hard engineering schemes – have an impact on natural processes. Groynes interrupt the movement of sediment by longshore drift, and sea walls can deflect high-energy waves along the coast. It is often the case that coastal defences protect one area but cause increased problems elsewhere along the coast. Some people consider these artificial structures to be ugly, ruining the natural landscape.

Figure 20 Coastal defences at Overstrand

Now test yourself

TESTED

1 Outline the impacts of geology on the geomorphic processes operating on the North Norfolk coast.
2 Use Figure 18 to identify the factors responsible for shaping these cliffs on the North Norfolk coast.

Exam practice

1 With reference to your chosen coastal case study, describe how geomorphic processes are affected by geology and climate.   [6]
2 Use Figure 20 to describe the coastal defences at Overstrand.   [4]
3 With reference to Figure 20 and your own knowledge about your chosen coastal case study, explain how coastal management works with geomorphic processes to affect the landscape.   [6]

ONLINE

Case study: A river landscape

In this section you need to study **one** river landscape. You should focus on the following aspects:

- geomorphic processes operating at different scales; how they are influenced by geology and climate
- landforms created by geomorphic processes
- the impact of human activity (including management) on geomorphic processes and the landscape.

Case study: River Wye

At over 210 kilometres in length, the River Wye is the fifth-longest river in the UK. From its source, high up in the Plynlimon Hills in central Wales, the River Wye flows roughly south-eastwards to join the River Severn at Chepstow. For much of its course the river flows through moorland and farmland. Considered to be a fairly natural river, there are few alterations resulting from human activity, for example the construction of dams and reservoirs. However, water power was used in paper production in the past, which was heavily polluting.

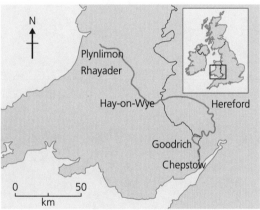

Figure 21 The course of the River Wye

Geomorphic processes

Geomorphic processes operate at different scales, both spatially (space) and temporally (time). The following list includes some of the geomorphic processes seen at the River Wye:

- Processes of river erosion, transportation and deposition are active along the course of the River Wye, producing a range of distinctive landforms, such as a V-shaped valley, meanders, floodplains and levees.

- Weathering and mass movement are active, particularly in the river's upper course in the Plynlimon Hills. Compared with river processes, these processes are more localised and take place at a smaller scale.

How are geomorphic processes affected by geology and climate?

Geology: For much of its upper course the River Wye flows across impermeable shales and gritstones. This accounts for the large number of tributaries that join the river. High rates of flow occurring after rainfall events enable the river to carry out significant erosion, forming steep-sided, V-shaped valleys (the upper slopes are actively weathered and affected by mass movement processes), waterfalls and rapids.

Away from the Plynlimon Hills, geology continues to influence geomorphic processes and landform development:

- Near Rhayader, alternating bands of hard and soft rock result in a series of rapids, a very popular site for canoeists.
- To the south of Hereford, weak mudstones and sandstones have been easily eroded to form an extensive flat valley characterised by sweeping meanders.
- In the south, between Goodrich and Chepstow, the river cuts through tough carboniferous limestone, forming an impressive gorge known as the Wye Valley.

Climate: The average annual rainfall of the River Wye basin is 725 mm. In the Plynlimon Hills, the annual rainfall can exceed 2500 mm. Much of this rainfall occurs in the winter when there is little growing vegetation to absorb the surplus water. This leads to rapid river flows, high rates of erosion and the potential for flooding.

Winter temperatures can be low, particularly in the Plynlimon Hills. This results in active freeze–thaw weathering on the exposed river valley sides. Active weathering, together with the high rainfall, promotes the processes of mass movement, such as sliding and slumping.

River landforms

The River Wye exhibits landforms of both erosion and deposition.

- V-shaped valleys are found in the upper course of the River Wye and along its tributaries in the Plynlimon Hills.
- Waterfalls (for example, Cleddon Falls) are formed on some of the River Wye's tributaries; the stretch of river near Rhayader is known for its spectacular series of rapids.
- The Wye Valley, a steep-sided river gorge, extends between Goodrich and Chepstow.
- Sweeping meanders have formed on the flat lowland plains to the south of Hereford.
- Levees and floodplains formed by extensive alluvium deposition are found in the middle and lower courses of the river.

Impact of human activity and management

The Environment Agency describes the River Wye as being 'highly urbanised'. The river flows through several large settlements including Rhayader, Hay-on-Wye, Hereford and Chepstow. Over 200,000 people live in the Wye and Usk (its tributary) river valleys. Much of the valley is used for farming, particularly in the river's middle and lower courses. The Wye Valley is popular with tourists who enjoy the attractive landscape, wildlife and opportunities for adventure tourism, such as kayaking, canoeing and climbing.

Flooding is a serious issue, particularly in urban settlements close to the river. To reduce the flood risk in Hereford the following steps have been taken:
- Storage lakes (such as Letton Lake) have been constructed above the town to store surplus water.
- Parts of the floodplain above the town are deliberately allowed to flood, relieving the pressure downstream.

- In Hereford itself, flood walls have been constructed to protect some 200 properties in the Belmont area at a cost of over £5 million (Figure 22).

Human activity and river management have affected geomorphic processes in the following ways:
- Tree planting in the river's upper course helps to stabilise the slopes, reducing mass movement. This reduces the amount of sediment in the river, which can choke the channel and increase the risk of flooding. With less sediment in the river, there will be a reduction in deposition further downstream.
- Fewer flooding events mean that there is less sediment available to construct floodplains, and natural levees might not build up. Artificial levees will have to be constructed instead.
- River banks along the course of the river have been stabilised by planting vegetation, improving access for anglers and walkers. Planting trees can reduce the height of floods by twenty per cent as they increase the amount of water that can be stored. The management of river banks will affect rates of flow and river processes, both in towns and in the countryside.

Exam practice

1 With reference to your chosen river case study, describe how geomorphic processes are affected by geology and climate. [6]
2 Look at Figure 22. How has the River Wye been managed to prevent flooding? [4]
3 With reference to your chosen river case study, explain how river management has influenced geomorphic processes in this landscape. [6]

ONLINE

Figure 22 **Flood defences protect property in Hereford, December 2012**

6 Why are natural ecosystems important?

The concept of an ecosystem

What are ecosystems?

REVISED

Ecosystems are natural areas in which plants, animals and other organisms interact with each other and the non-living elements of the environment. Plants are scientifically known as **flora** and animals are called **fauna**. The living elements of an ecosystem are known as **biotic**; the physical, non-living parts are known as **abiotic**.

Figure 1 Biotic and abiotic elements of an ecosystem

Biotic	Abiotic
Animals, including insects, birds and mammals Plants, including trees, grasses, mosses and algae, which provide food and shelter for animals Micro-organisms, such as fungi, which break down dead plants and animals, releasing nutrients into the ecosystem	Soils, which store water and carbon nutrients that plants can use Rocks, which help in the formation of soils; weathering releases nutrients stored in rocks into the ecosystem Sunshine and rain, which are needed for photosynthesis

Ecosystem: an area where living organisms and non-living elements interact with each other.

Flora: another term for the plants in an ecosystem.

Fauna: another term for the animals in an ecosystem.

Biotic: the living elements of an ecosystem.

Abiotic: the non-living elements of an ecosystem.

How are ecosystems 'interdependent'?

REVISED

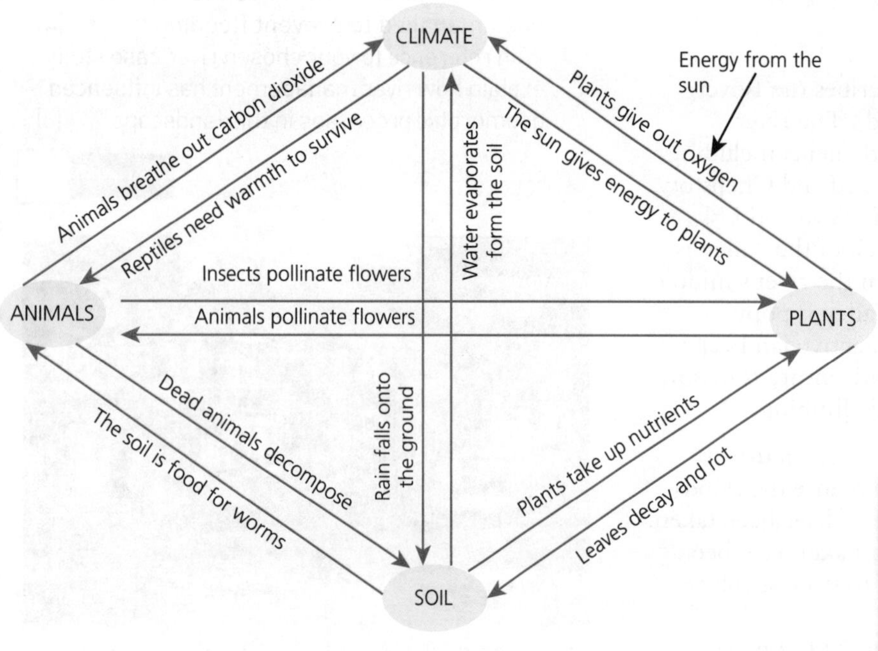

Figure 2 Interdependence in an ecosystem

Now test yourself

1. Explain how plants are dependent on the soil.
2. Explain how animals are dependent on the climate.
3. Explain how the climate is dependent on plants.

TESTED

Interdependence: the reliance of every form of life on other living things and on the natural resources in its environment, such as air, soil and water.

The global distribution and characteristics of biomes

What is the global distribution of the major biomes?

The world contains eight major **biomes**. They are mostly terrestrial (on land), but some, like coral reefs, are marine (in the sea). Each biome has unique climatic conditions that create distinctive environments for plants and animals to adapt to and survive.

> **Biome**: large-scale ecosystems that are spread across continents and have plants and animals that are unique to them.

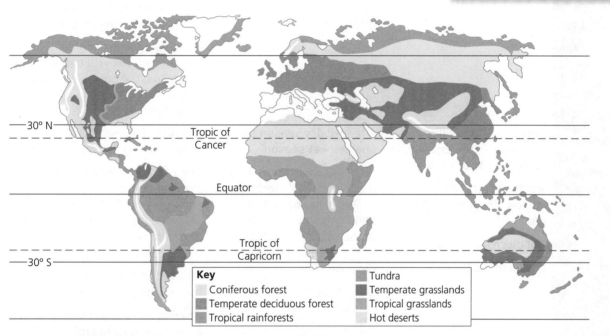

Figure 3 The global distribution of biomes

Key
- Coniferous forest
- Temperate deciduous forest
- Tropical rainforests
- Tundra
- Temperate grasslands
- Tropical grasslands
- Hot deserts

Exam practice

1 Which statement best explains the interdependence between soil and the climate? [1]
   (a) The Sun gives energy to plants.
   (b) Reptiles need warmth to survive.
   (c) Water evaporates from the soil.
   (d) Plants take up nutrients.
2 Describe how ecosystems are made up of both biotic and abiotic elements. [4]

What are the characteristics of the major biomes?

REVISED

Figure 4 Key characteristics of the major biomes

Biome	Location	Climate	Flora and fauna
Tropical rainforests	Around the Equator and within the tropics Amazon River Basin, South America South East Asia and Queensland, Australia	Hot and wet climate with no seasons Monthly temperatures are high throughout the year, between 26 °C and 28 °C Annual rainfall is often over 2000 mm; a thundery downpour happens most afternoons	15 million plant and animal species have been identified Vegetation exists in distinct layers, from the highest emergents to the canopy, under canopy and shrub layer Buttress roots support tall trees, and vine-like plants called lianas grow between trees Animals species include toucans, monkeys, chameleons and frogs; the poison dart frog is brightly coloured to warn off predators
Tropical grasslands (savannah)	Between 5° and 30° north and south of the Equator, in central parts of continents Most of central Africa, surrounding the Congo Basin Northern Australia Brazilian Highlands	A longer dry season and a shorter wet season Wet season arrives when the Sun moves overhead (ITCZ), bringing with it a band of heavy rain; 80% of the rain falls in the four or five months of the wet season During the dry season rainfall is as low as 10 mm Temperatures are high throughout the year but with a greater range than the rainforest; daily temperatures reach 25 °C	Tall and spiky pampas grass grows quickly to over 3 metres The baobab tree has large swollen stems and a trunk with a diameter of 10 metres to store water; it has a small number of leaves to reduce water loss via transpiration 40 species of hoofed animals, for example antelope Grazing species such as elephants, giraffe and wildebeest Carnivores, such as lions and hyenas, stalk herds
Hot deserts	Around 30° north and south of the Equator, typically on the west coast of continents around the tropics Sahara in northern Africa Mojave in North America	Daytime temperatures of 36 °C but temperatures at night can fall to −12 °C due to the lack of insulating cloud cover Annual precipitation is around 40 mm	Most plants are xerophytic (adapted to survive a lack of water) Cacti and yucca plants absorb water and have roots near the surface Camels have humps to store water and fat, and long eyelashes to keep the sand out Other species that have adapted to the extreme dryness are meerkats and sidewinder snakes

Figure 4 Key characteristics of the major biomes

Biome	Location	Climatev	Flora and fauna
Temperate grasslands	Between 40° and 60° north and south of the Equator, in central parts of continents Plains of North America Veldts of Africa Steppes of Eurasia	Summers are very hot, reaching over 38 °C, while winters are very cold, plummeting to as low as −40 °C Average rainfall varies from 250 mm to 750 mm, with 75% of rain falling in the summer season Melting snow helps the start of the growing season	Trees and shrubs struggle to grow, but some, such as willow and oak, grow in river valleys Tussock grasses grow to heights of 2 metres while buffalo and feather grasses grow more evenly to 50 cm. Burrowing animals such as gophers and rabbits, and large herbivores such as kangaroos and bison Carnivores including coyotes and wolves, as well as eagles
Temperate forests	Between 40° and 60° north and south of the Equator Eastern North America, Western Europe (including the UK) and New Zealand	Due to the tilt of the Earth and the angle of the Sun's rays, there are four distinct seasons Summers are warm and winters are mild Rainfall occurs throughout the year and ranges from 750 mm to 1500 mm, with an annual average temperature of 10 °C	Deciduous trees shed their leaves in the winter Oak trees reach heights of 30–40 metres; other tree species include ash, birch and hawthorn Grass and bracken grow on the forest floor Some animals migrate or hibernate in winter, for example black bears in North America Common species include squirrels, owls, pigeons, rabbits, deer and foxes

Now test yourself

TESTED

1 Give five examples of how plants and animals from any biome have adapted to survive in their environment.
2 Describe how temperature and rainfall vary between each of the biomes.
3 How does the position of the ITCZ affect patterns of rainfall in tropical grasslands?
4 State two places in which you would find:
   (a) hot deserts
   (b) tropical rainforests.
5 Why is there such a large range of temperatures across 24 hours in the hot desert?

Exam practice

1 Describe the pattern of temperature and rainfall in tropical rainforests. [4]
2 State two animals you would find in temperate forests. [2]
3 Describe the distribution of tropical grasslands. [3]

ONLINE

Polar regions

Location

Antarctica is a continent in the South Pole region covered by an immense ice shelf. The Arctic is in the North Pole and includes islands such as Greenland and the northern parts of countries such as Russia and Canada.

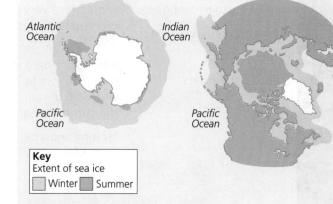

ANTARCTICA **ARCTIC**

Atlantic Ocean · Indian Ocean · Pacific Ocean · Pacific Ocean · Atlantic Ocean

Key
Extent of sea ice
☐ Winter ☐ Summer

Figure 5 Maps of the polar regions

Climate, flora and fauna

The climate of both regions consists of long, cold winters and short, cool summers. They are covered in snow and ice throughout the year, though the extent varies with the seasons (Figure 5). Temperatures rarely rise above freezing due to the low angle of the Sun in the sky. Due to the tilt of the Earth, each pole spends half of the year in darkness. Polar regions are dry, receiving only about 250 mm of rainfall per year.

Figure 6 Characteristics of the polar regions

	Antarctica (South)	Arctic (North)
Climate	South Pole winter temperatures vary from −62 °C to −55 °C	North Pole winter temperatures vary from −46 °C to −26 °C
	The weather in Antarctica is kept within the continent due to circumpolar winds travelling around the continent	The sea in the Arctic does not fall below −2 °C, causing the Arctic region to stay warmer than Antarctica
	Antarctica has an average altitude of 2300 metres (note that temperature decreases with altitude)	Relatively warm weather from the south travels north into the Arctic region via the Gulf Stream
Flora	Only 1% of the continent is ice free so plant life is less plentiful than in the Arctic	Approximately 1700 species of plant
	Around 100 species of moss and 300–400 species of lichen grow on exposed rocks	Mosses, grasses and alpine-like flowering plants
		Low shrubs, reaching heights of around two metres
		Treeless due to the permafrost
Fauna	Large numbers of penguin species such as gentoo, emperor and Adélie	Land mammals including polar bears, wolves, foxes and reindeer
	Fur seals and elephant seals	Sea mammals including walruses and whales
	Killer whales and minke whales	Some animals are able to migrate southwards during the winter months
	Both poles have rich seas due to large volumes of phytoplankton	

Now test yourself

TESTED

1 What are the similarities between the Arctic and Antarctica?
2 Explain two reasons why Antarctica is colder than the Arctic.
3 Why do trees not grow in the Arctic?
4 Why are there more animals than plants in the poles?

Coral reefs

Location

Known as the 'rainforests of the oceans', coral reefs cover less than one per cent of the world's ocean but are home to an estimated 25 per cent of all marine life. Some estimates say there are up to 2 million species living in coral reefs and 400 species of fish alone.

In total, 109 countries have coral reefs in their waters. For coral reefs to grow, they need warm water all year round with a mean temperature of 18 °C, and the water needs to be clear and shallow (no deeper than 30 metres) to ensure that there is enough sunlight for photosynthesis.

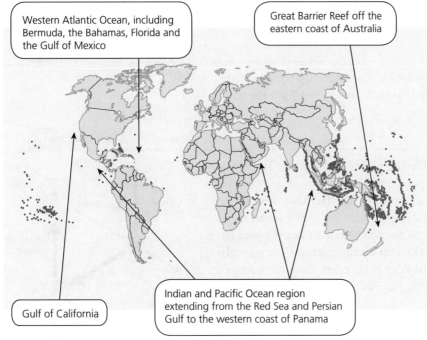

Western Atlantic Ocean, including Bermuda, the Bahamas, Florida and the Gulf of Mexico

Great Barrier Reef off the eastern coast of Australia

Gulf of California

Indian and Pacific Ocean region extending from the Red Sea and Persian Gulf to the western coast of Panama

> **Exam tip**
>
> For the Exam practice question, remember that 'describe' means that you must say what you see on the graph. You do not need to give any reasons for the climate. You would get one mark for describing the temperature and one for rainfall. Use data to support your point.

Figure 7 The global distribution of coral reefs

Flora

Sea grasses, such as turtle grass and manatee grass, are commonly found in the Caribbean Sea. Sea grasses are flowering plants that provide shelter and habitat for reef animals, and food for herbivores such as reef fish.

Fauna

Reefs are made of thousands of coral polyps. They look like plants but are, in fact, an animal related to the jellyfish. Each polyp is 2–3 cm in length and feeds on plankton. They make their own mineral skeleton from calcium carbonate. Other species found in coral reefs include parrot fish, starfish, clams, eels, dugongs, crustaceans and sponges.

Exam practice

Figure 8 Climate graph of Iqaluit, Canada (Arctic)

1 Using Figure 8, describe the climate of Iqaluit. [2]

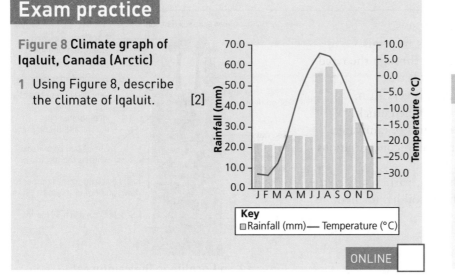

Key
Rainfall (mm) — Temperature (°C)

> **Now test yourself**
>
> 1 What environmental conditions are needed for coral to grow?
> 2 Name two places in the world where coral reefs are found.
> 3 What are coral reefs made of?

ONLINE

TESTED

7 Why should tropical rainforests matter to us?

The distinctive characteristics of tropical rainforests

Climate

REVISED

The climate is hot and wet with no seasons. The monthly temperatures range from 26 °C to 28 °C and annual rainfall can exceed 2,000 mm.

Nutrient cycle

REVISED

- **Biomass**: this is the total mass of living plants and animals and is the largest store of nutrients.
- **Litter**: this is the smallest store, explained by the fact that dead leaves decompose rapidly in the hot and damp conditions of the forest floor.
- **Soil**: the soil is also a small store. As soon as nutrients are released from the decayed litter layer, they are taken up by the nutrient-hungry plants and trees. This is why trees tend to keep their roots close to the surface of the ground, so they can take advantage of any nutrients as soon as they are available.

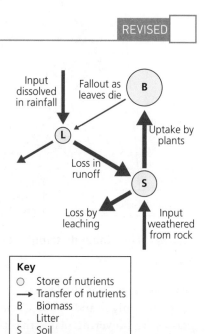

Key
- ○ Store of nutrients
- → Transfer of nutrients
- B Biomass
- L Litter
- S Soil

Figure 1 Nutrient cycling in the rainforest

Water cycle

REVISED

As the rainforest heats up during the morning, the water evaporates into the atmosphere and forms clouds to make rainfall. This is called convectional rainfall. Evapotranspiration is the process through which water is lost through pores in the leaves combined with evaporation. The canopy intercepts water on the leaves of trees. The removal of trees through deforestation can mean that there is less moisture in the atmosphere, causing rainfall to decrease. This can lead to droughts.

Soil profile

REVISED

- Soils in the rainforest are called latosols.
- They are generally shallow and lack minerals.
- The fertility of the soil is sustained through the rapid replacement of nutrients from dead leaves.
- The humus, or decomposed layer, is very thin. Minerals such as magnesium are leached quickly through the soil due to high amounts of rainwater.
- The top layers are red because of high concentrations of aluminium oxide and iron oxide.
- There is rapid chemical weathering, particularly of the original parent rock on which the soil has formed.

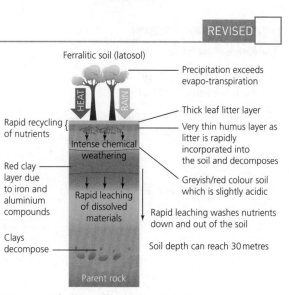

Figure 2 Soil profile in the rainforest

Indigenous people

Tropical rainforests are home to 50 million indigenous (native) people who depend on their surroundings for food, shelter and medicines. They are known as hunter-gatherers as they hunt animals and fish in the forest and gather wild fruits and nuts to eat. Today, most Amerindians use Western goods such as pots, pans and utensils, and make regular trips into towns to visit markets for trade. For centuries, farmers have practised a **sustainable** system called shifting cultivation (Figure 3).

> **Sustainable**: improving the current quality of life but still maintaining resources for the future.

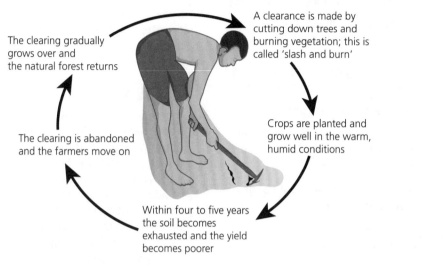

The clearing gradually grows over and the natural forest returns

A clearance is made by cutting down trees and burning vegetation; this is called 'slash and burn'

Crops are planted and grow well in the warm, humid conditions

The clearing is abandoned and the farmers move on

Within four to five years the soil becomes exhausted and the yield becomes poorer

Figure 3 Shifting cultivation

> ### Revision activity
>
> Using the model from Figure 2 in Chapter 6 on page 56, redraw it and add detailed annotations to show how interdependence works specifically in the tropical rainforest ecosystem.

Now test yourself

1 Name two minerals found in the soil of tropical rainforests.
2 What is the term for the type of rainfall in the rainforest?
3 Why is decomposition of the litter layer so rapid?
4 The soil is surprisingly infertile – why?

Exam practice

1 State two features of the tropical rainforest climate. [2]
2 Describe the cycle of nutrients in the tropical rainforest. [4]
3 Explain how indigenous people use the rainforest in a sustainable way. [4]

The value of tropical rainforests

Goods and services

Figure 4 Goods and services provided by rainforests

Goods	Services
Fruits and vegetables, for example passion fruit and bananas	Trees act as a carbon sink, storing carbon, as well as giving out oxygen
Nuts, for example brazil nuts and cashews	Trees reduce flood risk by increasing the time it takes for water to get into rivers by intercepting rain on their leaves and storing water in root systems
Flavourings, for example vanilla, cocoa and coffee	
Fibres for rattan, and bamboo furniture	Maintains fragile soils
Wood, for example teak, mahogany and rosewood	Source of income for indigenous people through agriculture and tourism
Rubber for car tyres	Provides a habitat for a range of flora and fauna
Medicines such as curare, which is used as a muscle relaxant in surgery, and quinine, which is an anti-malaria treatment	Maintains the water cycle, pumping moisture into the atmosphere

Goods: the items that we can physically extract from the rainforest to be used by people.

Services: the specific job, or role, that the ecosystem performs.

Exam tip

In a possible exam question about services, do not write about goods. Read the question carefully. Make sure you know the definition of *what* you are meant to write about as much as you understand the command word that tells you *how* to write your answer.

Deforestation

What is the scale of the problem?

Deforestation is the permanent removal of trees to be able to use the land for something else. Around 12.7 million hectares are being deforested around the world every year. Countries with significant rates of rainforest destruction include Brazil, Indonesia, Thailand and the Democratic Republic of the Congo.

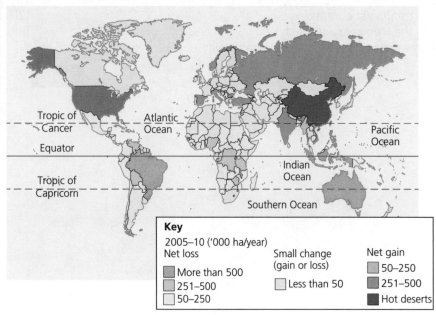

Key

2005–10 ('000 ha/year)

Net loss	Small change (gain or loss)	Net gain
More than 500	Less than 50	50–250
251–500		251–500
50–250		Hot deserts

Figure 5 Net changes in global forest areas

Exam tip

For a 4-mark question where you are required to use data, you need to:

● **Describe the pattern** using specific details from the map, including named regions of the world and compass directions. Two marks are awarded for this.

● Use information from the key to **add data (numbers)** to your descriptions. One mark is awarded for your use of data.

● The final mark is awarded for **communication skills**. Think carefully about your answer before you start writing to ensure that you get the mark for the order and clarity of your written communication.

Human activities in tropical rainforests

Logging

REVISED ☐

- Timber is harvested to make commercial products, such as furniture, paper and doors.
- Half of the wood is used for fuelwood.
- Logging companies argue that they cut down trees selectively, meaning they only take the trees they want. However, when one tree is cut down, many more are damaged.
- **Example**: Indonesia has one-tenth of the remaining rainforest in the world but is destroying it at a faster rate than any other country. The government has licensed much of the rainforest for logging to help develop its economy. Illegal logging has been responsible for the loss of 10 million hectares of forest. Through the burning of land, Indonesia accounts for eight per cent of global carbon emissions even though it covers barely one per cent of the Earth's land.

Cattle ranching

REVISED ☐

- Cattle ranching has become popular due to its relatively low risk and low maintenance in comparison to cash crops such as palm oil.
- It is not as vulnerable to changes in global prices or changes in climatic conditions.
- Due to the high demand for meat in the USA, farmers make a good profit.
- It can cause river siltation and soil erosion due to large areas of exposed land.
- **Example**: Around 80 per cent of Brazil's deforested area is now used for cattle, making it the largest cattle field in the world. It takes up 8.4 million hectares of land. The growth of the industry has been fuelled by new slaughterhouses and its proximity to the main market: the USA.

> **Cash crop**: crop grown and produced to be sold for profit.
>
> **Biofuel**: a fuel that comes from living matter, such as plant material.

Palm oil plantations

REVISED ☐

- Palm oil is one of the most profitable **cash crops** for developing countries. It is found in about half of all products sold in supermarkets, from margarines to shampoo.
- It is also used to create **biofuels**.
- Large areas need to be cleared by cutting down trees and burning them to release ash into the soil for short-term nourishment. Once the palm oil has been grown and harvested, the vast empty land is useless as it is infertile.
- **Example**: the largest palm oil plantations are in Indonesia, providing employment to 3 million people.

Figure 6 Palm oil plantation in Indonesia

Mining

- Gold, copper, diamonds and other precious metals and gemstones are found in rainforests around the world.
- Areas are also being surveyed for their oil and gas reserves.
- The mineral wealth of countries enables them to spend money on infrastructure projects such as electricity and roads.
- Environmental problems include soil and water contamination, particularly downstream of a mine due to toxic runoff into rivers.
- Communities are displaced from their land to make way for large machinery to gain access and products to be transported.
- **Example**: Peru is South America's top producer of silver and second-largest producer of zinc.

> **Exam tip**
>
> It is useful to be able to provide evidence for points that you make in the exam with *examples* from places around the world, particularly for questions worth 4, 6 or 8 marks.

Dam building

- Dams cause people to be moved from their land as vast areas of rainforest are flooded upstream.
- The dams are used to generate hydroelectric power (HEP) for other human activities such as logging.
- Dams interrupt natural river systems, holding silt back, which reduces the supply of nutrients for small-scale agriculture downstream.
- **Example**: more than twelve dams have been planned in the Malaysian state of Sarawak in Borneo; dam systems already exist along some of the biggest rivers in the world such as the Mekong in Laos.

> **Revision activity**
>
> Practise your skills of making geographical decisions. Rank human activities in tropical rainforests in order from most to least damaging. Justify your decisions.

Road construction

- It has been argued that large-scale road building projects are vital to help the poorest communities gain a better standard of living.
- Most roads, however, are built to make mines and HEP stations possible.
- **Example**: a road-building project in the Congo rainforest has led to 50,000 kilometres of road being constructed. Although this is a legal project led by the government, it has resulted in the creation of further illegal roads crossing the main road to enable logging to take place.

Figure 7 A new tarmac road being laid through the Congo forest

Now test yourself

1 Why do logging companies argue that their industry is not as damaging as others?
2 Why is cattle ranching common in South American countries?
3 Why are palm oil plantations so destructive to the environment?

Exam practice

Evaluate the level of exploitation caused by contrasting human activities in the rainforest. [8]

ONLINE

> **Exam tip**
>
> In the exam you will have 8–10 minutes to answer an 8-mark **evaluate** question. Exam practice question 1 should be answered in approximately three paragraphs. Follow a simple structure. Paragraph 1: 'On the one hand...', paragraph 2: 'However...', paragraph 3: 'In conclusion...'.

Case study: Sustainable management of an area of tropical rainforest

Case study: Crocker Range Biosphere Reserve, Borneo

Location

Crocker Range Biosphere Reserve, in northern Borneo, was designated by UNESCO in 2014. It stretches 120 kilometres from north to south and 40 kilometres from east to west. The reserve has 350,585 hectares of various altitudes up to 1500 metres. The highest temperature is 32°C, and it has an annual rainfall of 3000 mm.

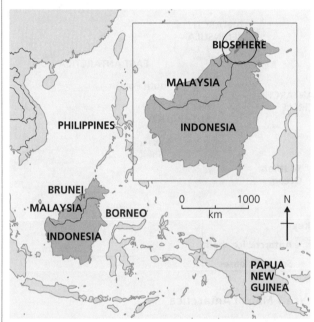

Figure 8 Location of Crocker Range Biosphere Reserve

How is it being protected?

Core area (144,492 hectares): there are six permanent plots (50 m × 50 m) for ecological monitoring. The core area is strictly for long-term research programmes, environmental education and tourism. Its estimated population is a mere 200 people. There are around 30 families who engage in agricultural and natural resource use, producing rubber and cocoa, as well as working with local authorities to protect the reserve. Livelihoods are mostly sustained through agricultural activities such as hill paddy, coconut and fruit farming, covering approximately 160 hectares.

Why does it need protecting?

The reserve itself has been identified as a region with a wide variety of plant and animal species due to its varying height above sea level. Many of those are endemic, which means that they can only be found in this part of the world.

- In terms of fauna, the number of species recorded in the whole of the Crocker Range Biosphere Reserve and its surrounding area includes: 101 mammals, 259 birds, 47 reptiles, 63 amphibians and 42 freshwater fish.
- Two endemic *Rafflesia* species are found in the Crocker Range, with a total of 737 plant species recorded in the biosphere reserve's eastern area.
- The Crocker Range is also the habitat of several endangered species, such as orang-utans, sun bears and clouded leopards.

Exam practice

Evaluate the success of **one** attempt to sustainably manage an area of tropical rainforest. [6]

ONLINE

Buffer zone (60,313 hectares): in the buffer zone, where a total of 52 villages are confirmed, small-scale agriculture and rubber tree plantation are the main human activities.

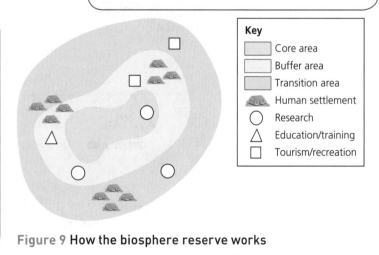

Figure 9 How the biosphere reserve works

Transition area (145,779 hectares): the transition zone features at least 264 villages. Communities engage in subsistence agriculture and small-scale plantations, producing vegetables and other products for Kota Kinabalu city.

8 Is there more to polar environments than ice?

The distinctive characteristics of the Antarctic and the Arctic

The Antarctic and the Arctic

REVISED

The Antarctic is a huge continental landmass located almost entirely within the Antarctic Circle (Figure 1). The South Pole is at its centre. Covering an area of over 13 million square kilometres, it is 60 times the size of the UK and larger than the whole of Europe. It is one of the seven continents of the world. Almost the entire continent is covered by ice, which is several kilometres thick in places. With no permanent inhabitants, Antarctica is known as the world's last great wilderness.

In contrast, the Arctic region covers land areas within the Arctic Circle and the Arctic Ocean, which in the winter is largely frozen (Figure 2). It is important to remember that the Arctic ice is frozen sea ice; there is no continent as is the case with Antarctica. The Arctic land areas include large areas of northern Russia, Scandinavia, Greenland, Canada and northern Alaska (USA). There are numerous islands within the Arctic. An estimated 5 million people live and work in the Arctic.

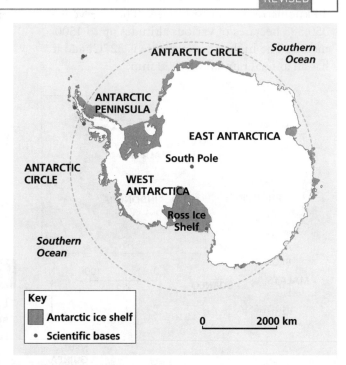

Key
- Antarctic ice shelf
- Scientific bases

0   2000 km

Figure 1 Map of Antarctica

Figure 2 Map of the Arctic

Figure 3 identifies some of the distinctive characteristics of the Antarctic and the Arctic.

Figure 3 Distinctive characteristics of the Antarctic and the Arctic

	Antarctica	Arctic
Climate	Extreme polar climate with winter temperatures dropping well below −50°C and summer temperatures well below freezing. Due to the polar high pressure centred at the South Pole, dry air travels outwards from the centre of the continent. This results in the low precipitation of this 'dry desert'. Antarctica experiences extremely strong winds.	The Arctic also experiences a polar climate but with much greater variations in temperature. Winter temperatures drop to below −30°C, whereas summer temperatures may rise to 2–12°C in places. As with the Antarctic, much of the Arctic experiences low precipitation totals. At the coastal borders, however, heavy snowfall can occur during the winter as relatively warm moist air is blown in from the sea.
Features of land and sea	Some 99% of the continent is covered by ice with the remainder being bare rock. The 5000-metre high Transantarctic Mountains stretch for 3500 kilometres through the centre of the continent, dividing west Antarctica from east Antarctica. Ice sheets cover the mountains, although some individual peaks (called nunataks) stick out through the ice. Several vast ice shelves extend into the sea, for example the Ross Ice Shelf (Figure 1).	Much of Greenland is covered by an ice sheet, the second-largest ice sheet after Antarctica. Some glaciers flow into the sea where they calve to form icebergs. Much of the rest of the land fringing the Arctic experiences permanently frozen ground – permafrost. During the summer, the surface few centimetres melt to create a waterlogged, boggy landscape with many ponds, lakes and depressions. Sea ice develops during the winter months, although its extent has diminished in recent years due to global warming.
Flora and fauna	Few plants can survive the extremely hostile conditions (extreme cold, strong winds, dryness (aridity), lack of soil). Apart from two species of grass, only algae and lichens can survive. Penguins nest on the ice shelves but most marine creatures live around the edges of the continent. The cold waters of Antarctica are rich sources of food for fish and mammals.	There are large areas of coniferous trees in the far south. Further north and into the Arctic Circle, trees are replaced with tundra, low-growing cushions of vegetation and stunted shrubs and trees. Plants have adapted to the extremely cold and dry conditions by developing small waxy leaves and growing close to the ground to shelter from the winds. The tundra is rich in wildlife, particularly insects and small mammals. Larger animals, such as the Arctic fox, bison and reindeer, can survive the extreme cold by growing thick coats of insulating fur. In marine environments, seals, whales and walruses are common. Polar bears are the iconic threatened predator found on ice sheets and coastal regions.

Now test yourself

TESTED ☐

What are the distinctive characteristics of the Antarctic?

Revision activity

Use a Venn diagram to identify the similarities and differences between the Antarctic and the Arctic.

How do ecosystems in the Antarctic and Arctic demonstrate interdependence?

In this section you need to focus on **one** polar region: either the Antarctic or the Arctic.

Ecosystems show clearly the interdependence between biotic (living) factors such as plants, animals and people, and abiotic (non-living) factors such as climate, water and soil. These extreme environments create challenging conditions for ecosystems:

- The lack of liquid water limits the growth of vegetation in the Arctic.
- The extreme climate and lack of soil accounts for Antarctica being the least biodiverse region in the world, with very few species of plant and no resident land animals.
- Marine ecosystems are highly developed and diverse in both the Antarctic (Figure 4) and the Arctic (Figure 5).
- Biodiversity is greater in the Arctic than in the Antarctic. The Arctic climate is less extreme and a soil has usually developed. There are more species

of plants and animals, and they have become well adapted to the conditions. Plants are low growing to prevent wind damage and have small waxy leaves to retain moisture. Animals such as polar bears have thick, insulating fur, small ears (to reduce heat loss) and insulated foot pads.

- Deep snow in the Arctic provides cover for small rodents and mammals.
- While there are no permanent settlements in the Antarctic, there are scientific research stations making the most of the pristine conditions to study environmental change. The Arctic is used by indigenous people for fishing, hunting and reindeer herding.
- Fragile polar ecosystems are at risk of damage by people, particularly through mineral exploitation (for example, oil) and tourism. Tourism is growing rapidly in both the Antarctic and the Arctic.

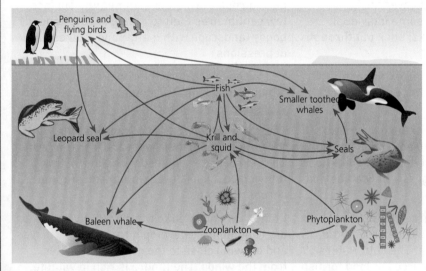

Figure 4 A typical Antarctic food web

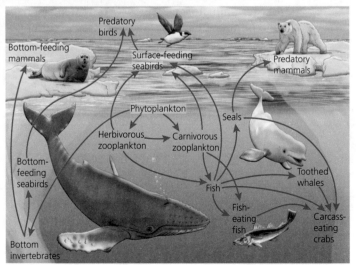

Figure 5 A typical Arctic food web

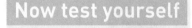

Now test yourself

Suggest reasons why the Arctic shows greater biodiversity than the Antarctic.

TESTED

Revision activity

Construct a spider diagram focusing on interdependence in **either** the Antarctic **or** the Arctic.

The impacts of human activity on ecosystems in the Antarctic and the Arctic

In this section you need to study the impacts of human activity on ecosystems in **either** the Antarctic **or** the Arctic.

Indigenous people (Arctic)

Groups of indigenous people have inhabited parts of the Arctic for thousands of years. Well adapted to life in the extreme cold, these tightly-knit communities have survived through hunting, fishing and herding animals such as reindeer in northern Scandinavia.

Indigenous people have a deep knowledge and understanding of their natural environment; they respect nature and live a sustainable lifestyle. When hunting, they take only what they need and they use every part of an animal or fish, wasting nothing.

Scientific research (Antarctic)

The pristine conditions associated with the world's 'last great wilderness' mean that the Antarctic provides unique opportunities for studying weather and climate, glaciology, marine biology, space and geology.

Twenty countries operate 40 year-round research stations in Antarctica. Today the focus for much of the scientific research is global climate change, involving the extraction of deep ice cores to examine past climates.

In the past, the operation of research stations has had serious impacts on ecosystems:
- sewage, solid waste and oil drums caused pollution
- alien species, such as non-native spiders, mosses and fruit flies, now inhabit the USA's McMurdo Station; this can cause an imbalance in the ecosystem
- aerial surveys have suggested that communication and energy production at research stations may be affecting magnetic fields. This could have an impact on plants, animals and even humans within a radius of a few hundred metres. For example, it could affect blood flow, the activity of the pineal gland (which regulates wake/sleep functions and seasonal patterns) and the control of pain in humans and animals.

Today there are strict environmental guidelines designed to minimise the impact on marine and terrestrial ecosystems. The stations operate modern sewage treatment works. Recycling is mandatory at US research stations, and solid waste is shipped outside Antarctica.

Now test yourself

Outline some of the harmful impacts of scientific research on Antarctic ecosystems.

Tourism (Antarctica)

Tourism has grown rapidly in both the Arctic and the Antarctic. Improvements in technology (for example, in transport and clothing) have made these extreme and remote environments more accessible. Increased life expectancy and higher disposable incomes mean that more people can afford to visit these expensive destinations.

Some 25,000–30,000 people visit Antarctica each year and numbers are expected to grow in the future. Tourism has had some negative impacts on the ecosystem, including:

● marine pollution caused by discharges and waste from ships
● pollution and erosion at landing sites, which can become congested at peak times
● intrusive behaviour (for example, photography), particularly around penguins
● ships can occasionally be damaged or even sink due to the presence of pack ice.

Mineral extraction (Arctic)

The Arctic contains rich mineral resources, including various metals, oil and gas. Oil is transported south via the Trans-Alaska Pipeline. In constructing the pipeline, some sections were raised high above the ground to allow the annual migration of caribou to take place without interference. Despite the success of the Trans-Alaska Pipeline, oil spills have occurred, causing considerable damage to the environment:

● In 1989, the oil tanker *Exxon Valdez* ran aground on rocks in Prince William Sound, discharging vast quantities of oil into coastal waters. The oil caused considerable damage to marine and coastal ecosystems, killing thousands of birds and animals (Figure 6).
● Oil spills resulting from ruptured, poorly-maintained pipelines in Siberia, Russia, have polluted rivers, lakes and woodlands, causing considerable damage to local ecosystems.

> ### Now test yourself
>
> Describe the impacts that mineral extraction can have on polar ecosystems.
>
> TESTED

Figure 6 Dead Californian whale resulting from the *Exxon Valdez* oil spill

Fishing (Arctic)

REVISED

The cold, nutrient-rich Arctic waters are extremely important fishing grounds. About 70 per cent of the world's white fish is caught in the Arctic Ocean. Many coastal communities throughout the region depend on fish as a source of food and income. Fishing has an enormous impact on the ecosystem:

- Overfishing and illegal fishing using crude dragnet techniques can unbalance coastal ecosystems and devastate future fish stocks.

- In the past, Newfoundland and Labrador overfished part of the Arctic Ocean leading to a collapse in fish stocks.
- There is concern about the possible impacts of climate change on sea temperatures, which will affect Arctic marine ecosystems and could lead to the disappearance of important fisheries.

Now test yourself

TESTED

Describe how fishing can affect Arctic ecosystems.

Whaling (Antarctic)

REVISED

Many countries were involved in the whaling industry in the Southern Ocean in the past. By the late 1930s, approximately 50,000 whales were being killed annually, posing a considerable threat to whale stocks.

Commercial whaling officially ended in 1986, although Iceland and Norway are still engaged in limited whaling. Japan issues permits for 'scientific whaling' in Antarctic waters, although in 2014 the International Court of Justice declared that this was just a front for commercial whaling.

Whales are vital to the food chain and are important in maintaining a healthy ocean ecosystem. The decline of stocks in the Southern Ocean has had an impact on the marine ecosystem:

- A whale can eat up to 40 million krill each day – if whale numbers decrease, the massive krill surplus unbalances the food chain.
- Dead whales that sink to the seabed are important sources of food for other organisms.
- Whale faeces contain nutrients that stimulate the growth of phytoplankton, which is a major source of food for fish and other marine organisms. Phytoplankton absorbs large quantities of carbon from the atmosphere, helping to regulate the world's climate.
- Whales are the top predators in the marine ecosystem. A reduction in numbers due to hunting will have a huge impact on the number of species lower down the food chain and will unbalance the ecosystem (Figure 7).

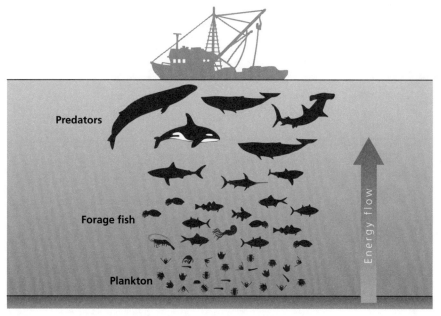

Figure 7 How whaling affects the marine ecosystem

Revision activity

Draw a spider diagram to give an overview of the impacts of human activity on ecosystems in polar environments. Use colour coding to identify Arctic and Antarctic examples.

8 Is there more to polar environments than ice?

Exam practice

1 Outline the distinctive climatic characteristics of the Antarctic and the Arctic. [4]
2 Explain the differences in flora and fauna between the Antarctic and the Arctic. [4]
3 For **either** the Antarctic **or** the Arctic, describe the interdependence that exists within ecosystems. [6]
4 With reference to either the Arctic or the Antarctic, describe the impacts of tourism on ecosystems. [4]
5 With the help of Figure 7 and your own knowledge, explain how whaling can affect **either** the Arctic **or** the Antarctic marine ecosystem. [6]
6 With the help of Figure 5, outline the impacts of oil spills on Arctic ecosystems. [4]

ONLINE

Exam tip

When describing human activities in polar environments, ensure that you focus clearly on the **impacts** to the ecosystems.

Learn case study information for either the Arctic or the Antarctic – make sure that you know which case study you are using for each human activity.

Case study: Small-scale sustainable management in the Antarctic and the Arctic

In this section you need to focus on a case study to examine **one** small-scale example of sustainable management in **either** the Antarctic **or** the Arctic. We will consider sustainable tourism in Antarctica.

> **Sustainable management**: management that protects the environment and meets the needs of present and future generations.

Case study: Sustainable tourism management at Union Glacier, Antarctica

Sustainable management involves striking a balance between providing access for tourists and protecting the environment for future generations.

Union Glacier is a large expanse of ice in the Ellsworth Range, part of the Transantarctic Mountains that stretch through the centre of the continent (Figure 8). Operated by Antarctic Logistics & Expeditions (ALE), Union Glacier serves as a logistics hub for private expeditions, research and limited tourism.

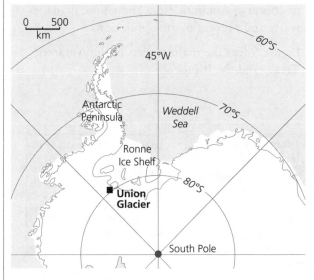

Figure 8 Location of Union Glacier

Each year a camp is built beside a natural blue-ice runway capable of handling cargo planes. The camp operates during the summer (November to January) and is dismantled each year at the end of the season.

Up to 70 guests can be housed in sturdy clam tents. There is a large dining tent, a fully fitted kitchen, and communal toilet and shower facilities.

Guests visiting the camp can take part in a number of activities:
- walking or cross-country skiing, for example in the footsteps of the Norwegian explorer Amundsen up the Axel Heiberg Glacier
- climbing, including a chance to climb Mount Vinson
- visiting penguin colonies – strict rules apply, such as groups being limited to twenty people and no-one being allowed to approach penguins closer than five metres.

To minimise the impact on the environment, the camp follows strict guidelines and internationally agreed protocols:
- strict biosecurity measures to prevent alien species being introduced by contaminated footwear and equipment
- guests are encouraged to use a minimum amount of water when showering and to do so infrequently to cut down on 'grey' water (relatively clean domestic wastewater, excluding sewage)
- human waste is separated for ease of disposal and transported away from the camp; guests are provided with 'pee bottles'
- tents are naturally heated by the 24-hour sunlight in the Antarctic summer
- some equipment is powered by solar energy to reduce the use of diesel.

Exam practice

With reference to your chosen case study, examine one small-scale example of sustainable management in **either** the Antarctic **or** the Arctic. [6]

ONLINE

Case study: Global sustainable management in the Antarctic and the Arctic

In this section you need to focus on a case study to examine **one** global example of sustainable management in **either** the Antarctic **or** the Arctic. We will consider the **Antarctic Treaty**.

Antarctic Treaty: agreed in 1961 to help regulate human activity in the area, and resolve political disputes over territory.

Case study: The Antarctic Treaty

The Antarctic Treaty is one of the most successful international agreements of all time. It came into force in 1961, initially signed by twelve countries that had been actively involved in scientific research in the late 1950s.

Today, 52 countries (comprising over 80 per cent of the world's population) have signed the treaty, demonstrating their commitment to retaining this vast untouched wilderness for the benefit of all humanity (Figure 9).

Twenty-eight nations form a consultative body (the Consultative Parties), which meets annually to consider current issues and make recommendations to promote sustainable management. In addition to the Antarctic Treaty, three international agreements have been reached:

- Convention for the Conservation of Antarctic Seals (1972)
- Convention on the Conservation of Antarctic Marine Living Resources (1980)
- Protocol on Environmental Protection to the Antarctic Treaty (1991).

The Protocol on Environmental Protection came into force in 1998. It designates all of the area south of 60°S line of latitude as 'a natural reserve, dedicated to peace and science'. In outlining basic principles for human activity, it stresses the need to conserve Antarctic flora and fauna, manage waste and prevent marine pollution.

Figure 9 lists the main aspects of the Antarctic Treaty and the Protocol.

Sets aside the potential for sovereignty disputes between Treaty parties.

Promotes international scientific research and co-operation including the exchange of research plans and personnel, and requires that results of research be made freely available.

Ensures that Antarctica should be used exclusively for peaceful purposes – military activities, such as the establishment of military bases or weapons testing, are specifically prohibited.

Prohibits nuclear explosions and the disposal of radioactive waste.

Preserves historic sites, such as the huts used by explorers Scott and Shackleton.

Puts in place a dispute settlement procedure and a mechanism by which the Treaty can be modified.

Requires parties to give advance notice of their expeditions and promotes logistical co-operation and communication to aid safety of all on the continent.

Protects the Antarctic environment and bans any activity relating to mineral resources.

Figure 9 The Antarctic Treaty and Protocol: global sustainable management

Exam practice

With reference to your chosen case study, evaluate the success of one global example of sustainable management in **either** the Antarctic **or** the Arctic.   [6]

ONLINE

Revision activity

Construct a revision diagram, such as a spider diagram, to summarise the key aspects of the Antarctic Treaty and the Protocol on Environmental Protection.

9 Why do more than half the world's population live in urban areas?

The global pattern of urbanisation

In 2007, the UN announced that more than half of the world's population is now living in urban (city) areas. The increase in the numbers of people living in urban areas is called **urbanisation**. Globally, the number of city dwellers increases by an estimated 180,000 people every day. By 2050, 75 per cent of the world's population could be living in towns and cities.

> **Urbanisation**: the process of towns and cities developing and becoming bigger as their population increases.

Urban growth rates

REVISED

The pattern of growth has not been the same everywhere. Figure 1 describes the general patterns of growth in advanced countries (ACs), emerging and developing countries (EDCs) and low-income developing countries (LIDCs).

Figure 1 Growth in ACs, EDCs and LIDCs

ACs	EDCs and LIDCs
Cities in Europe and North America reached the peak of their growth in the 1950s or earlier.	Cities in Asia and Africa have now overtaken the earlier cities of Europe and North America.
The most sustained period of growth took place during the Industrial Revolution in the late 1700s to 1800s.	Economic development in urban areas has driven rural–urban migration, causing younger people in rural areas to move to urban areas in search of jobs.
The 'baby boom' following the Second World War and the building of new houses led to urban sprawl and the growth of cities.	Many of these people then have children in the city, leading to high rates of natural growth.
London and Paris were the first 'millionaire' cities (population of 1 million).	Almost 200 million people moved to urban areas between 2000 and 2010.
Most ACs now have populations that are more than 70 per cent urban.	Currently, the highest levels of growth are seen in cities such as Dhaka (7 per cent), Lagos (5.6 per cent) and Delhi (4.6 per cent).

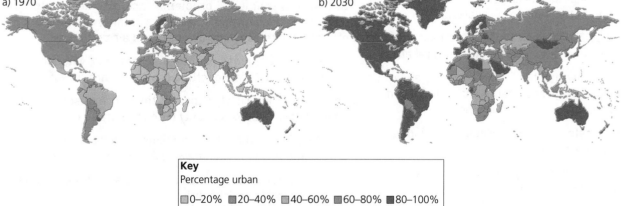

a) 1970   b) 2030

Key
Percentage urban

☐ 0–20% ☐ 20–40% ☐ 40–60% ☐ 60–80% ■ 80–100%

Figure 2 The predicted percentage urban population of countries in 2030

World cities and megacities
World cities

Headquarters of multinational companies

Cultural opportunities, for example live music, cinema and opera

Centre for innovation in business

Highly rated universities, often specialising in research

World city

Centre for media and communications

Home to an important stock exchange or major banks

Integration in the global economy

Important port facilities

Major centre for manufacturing

Figure 3 Common characteristics of world cities

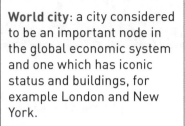

World city: a city considered to be an important node in the global economic system and one which has iconic status and buildings, for example London and New York.

Megacity: usually defined as a city that has a population of over 10 million, although the exact number varies.

Megacities

The rapid rate of growth in urban areas has led to the creation of a number of cities with populations of over 10 million people, known as megacities. They are often, though not always, capital cities.

Advantages and disadvantages of megacities in LIDCs

LIDCs have the greatest potential for growth, with around 2 million people a week moving into cities in Africa and Asia. This creates a range of advantages and disadvantages.

Informal sector: refers to jobs that don't offer regular contracted hours, salary, pensions or other features of more formal employment.

Figure 4 Advantages and disadvantages of megacities in LIDCs

Advantages	Disadvantages
Allows industry and finance to cluster together and take advantage of a ready market, labour force and access to markets abroad	Inward migration tends to happen quicker than the pace of economic and social development in the megacity
Provision of education and basic infrastructure (roads, water, electricity) is often better than in rural areas	Universities and health centres are usually in wealthier areas, so inaccessible to most of the population
Cities usually have lower infant mortality rates and higher life expectancy than rural areas	Serious environmental problems, including local water shortages, widespread subsidence of land, contamination of groundwater, sewage disposal and air pollution
The growth of the **informal sector** allows local entrepreneurial talent to thrive and helps to tackle unemployment	As many of the megacities are on coastlines, valuable habitats and natural coastline protection are threatened
Self-help housing provides a solution to housing shortages	The government rarely supports the informal sector as it does not provide tax and only helps the urban poor
Strong community and employment networks in slums	People living in slums tend to occupy land that is not fit for development, so they can be at risk of landslides and floods
	Lack of co-ordination across administrative areas makes the planning of services difficult

Changing distribution of megacities since 1950

In the 1950s there were only two megacities, but there are now over 30. Some of the cities have grown to populations of over 20 million people. Although the number of megacities is growing, they still have only five per cent of the world's population at present.

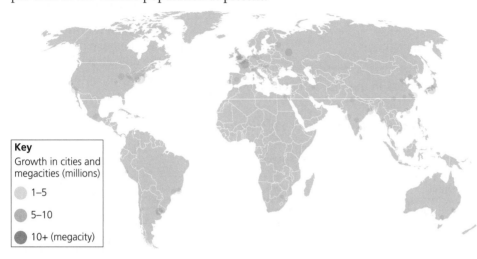

	Top 15 cities in 1950 (millions)
1	New York-Newark, USA (12)
2	Tokyo, Japan (11)
3	London, UK (8)
4	Paris, France (6.5)
5	Shanghai, China (6)
6	Moscow, Russia (5)
7	Buenos Aires, Argentina (5)
8	Chicago, USA (5)
9	Calcutta, India (4.5)
10	Beijing, China (4)
11	Osaka-Kobe, Japan (4)
12	Los Angeles, USA (4)
13	Berlin, Germany (3)
14	Philadelphia, USA (3)
15	Rio de Janeiro, Brazil (3)

Key
Growth in cities and megacities (millions)

- 1–5
- 5–10
- 10+ (megacity)

Figure 5 Global distribution of major cities and megacities in 1950

	Top 15 cities in 2014 (millions)
1	Tokyo (38)
2	Delhi (25)
3	Shanghai (23)
4	Mexico City (21)
5	São Paulo (21)
6	Mumbai (21)
7	Osaka-Kobe (20)
8	Beijing (19)
9	New York-Newark (19)
10	Cairo (18)
11	Dhaka, Bangladesh (17)
12	Karachi, Pakistan (16)
13	Buenos Aires (15)
14	Calcutta (15)
15	Istanbul, Turkey (14)

Key
Growth in cities and megacities (millions)

- 1–5
- 5–10
- 10+ (megacity)

Figure 6 Global distribution of major cities and megacities in 2014

Exam practice

1 Use Figure 2 on page 77 to complete the following sentence: 'The population of Brazil is predicted to be ... urban in 2030.' [1]
2 Explain the challenges and opportunities faced by megacities. [4]
3 Describe the characteristics of a world city, such as London. [4]
4 Describe the difference between the distribution of megacities in 1950 and 2014. [3]

ONLINE

Rapid urbanisation in cities

Causes of rapid urbanisation in LIDCs

LIDC cities are growing the fastest in the world. The highest population growth rate is in Africa, where urban populations are expected to triple in size by 2050. This increase is due to both **rural to urban migration** and **internal growth**.

> **Rural to urban migration**: people moving from rural areas to live in the cities.
>
> **Internal growth**: when people who have moved into the cities have lots of children.

Push and pull factors in migration

> **Push factor**: a negative factor that results in the movement of people away from an urban/rural area.
>
> **Pull factor**: a positive factor that attracts people into an urban/rural area.

Push factors	Pull factors
Few services, such as education and health care	Greater range of employment with higher wages
Wages are at poverty levels in many countries	Better health care systems and schools
Lack of job opportunities – jobs tend to be limited to agricultural work	Stories filter back to the villages of people doing better in the city, encouraging more to move
Poorer infrastructure	Political and religious freedom
Natural disasters, such as drought and flooding	Stable government
Poor electricity and power supplies	Better access to food
Crop failure	Better-quality housing
Lack of clean water	More entertainment – the 'bright lights' of the city
Increased food insecurity as more young people leave to live in the cities	More transport networks, such as road, rail and air

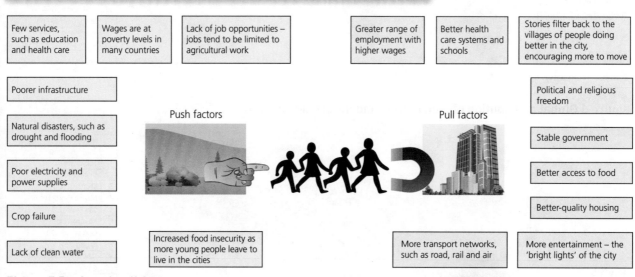

Figure 7 Push and pull factors

Internal growth

Once people have arrived in the city and found employment and housing, they tend to have children. This increase in birth rate can result in a rapid rate of population growth, particularly in LIDCs where there is a large, youthful population. ACs tend to have the opposite problem: an ageing population.

> **Exam tip**
>
> Students commonly mix up push and pull factors. Focus on the terms: push factors 'push' people out of the village; pull factors 'pull' people into the city.

Now test yourself

1 What are the two causes of rapid population growth in LIDC cities?
2 What are the environmental push factors from villages?
3 What are the economic pull factors into the cities?

Consequences of rapid urban growth in LIDCs

REVISED

Informal sector

The informal sector involves people finding their own employment. This includes jobs as beach vendors, shoe shiners, car washers and litter pickers. It is estimated that over 50 per cent of jobs in Mexico are in the informal sector, and up to 80 per cent of jobs in India.

Jobs in the informal sector require little capital to set up, require few skills, are labour intensive and small scale. People working in this sector do not pay taxes and therefore do not contribute directly to the country's gross national product (GNP). These workers do not have any legal rights and would not receive advantages such as holiday and sick pay.

Informal housing

Informal housing, such as slums or squatter settlements, is built on land that does not belong to the people building on it. It is usually land that is unsuitable for building, such as:

- a dry river bed, which can fill with water after rain
- land that is close to industrial activity, which could be bad for people's health
- steep and unstable slopes, which are vulnerable to landslides and flooding
- land either side of railway tracks.

Infrastructure is poor in areas of informal housing, and there are problems with the reliability of electricity and water supplies. Children may have to leave education and work in the informal sector to support their families. With high population densities, disease spreads easily and crime is common. Over time, areas can be improved through self-help schemes.

Figure 8 Slums and skyscrapers in Manila, Philippines

Exam practice

1 Describe the common locations for informal housing in LIDC cities. [3]
2 Explain the challenges faced by people living in informal housing. [4]

ONLINE

Urban trends in advanced countries

Suburbanisation

> **Suburbanisation**: a change in the nature of rural areas such that they start to resemble the suburbs.

Figure 9 Causes and consequences of suburbanisation

Causes	Consequences
Suburbanisation started in the mid- to late twentieth century, when public transport and private car ownership meant that commuters could live further out from the city centre. There was a move towards home ownership in the UK during the 1970s, which led to private housing estates being built on the edges of cities. Building in these areas allowed people to have more land for gardens and more public open space, compared with housing areas nearer the town centre.	Local shopping centres have been constructed, along with many primary schools and a smaller number of secondary schools. Created a demand for out-of-town retail parks. Buildings in the city are left vacant. These buildings might quickly start to look derelict and be vulnerable to vandalism, preventing inward investment in the city. Increased congestion and pollution from commuting.

Counter-urbanisation

> **Counter-urbanisation**: the movement of people from urban areas into rural areas; these may be people who originally made the move into a city.

Figure 10 Causes and consequences of counter-urbanisation

Causes	Consequences
People moving out of the city tend to be the most affluent and the most mobile. They tend to be those with young children who think that the countryside will be a better place to bring them up. The push factors from the city include traffic congestion, higher cost of living, perception of high crime, poor air quality, and the dream of the rural idyll. Better road and rail links to city centres have enabled people to live further away from their place of work and commute easily. Businesses with offices are now moving to more rural locations on the edge of cities where the land prices are cheaper and the quality of life for the workers can be better. With improvements in high-speed broadband in rural areas, new industries can locate anywhere. Similarly, improvements in telecommunications have enabled more people to work from home and still be in touch with their offices in cities.	**Counter-urbanisation** creates dormitory villages, where residents work in the city during the day and only return to the rural area in the evenings. House prices increase rapidly in relatively rural areas just outside urban centres, such as Hemel Hempstead outside of London. Long-term residents of villages fear that the character of these settlements is changing. Local people can be priced out of the market as wealthier city people buy up and renovate older properties, raising the profile of the settlement and subsequently the cost of properties. This process is called gentrification.

Re-urbanisation

> **Re-urbanisation**: the use of initiatives to counter problems of inner-city decline.

Figure 11 Causes and consequences of re-urbanisation

Causes	Consequences
People are returning to live in the city, particularly in inner-city areas.	The redevelopment of inner-city urban areas creates new jobs and homes, which attracts people from outside to move there.
Government initiatives encourage people and businesses back into the city; for example, staff may be paid a premium for new jobs in deprived areas.	There can be a lack of affordable housing, which can lead to expensive new apartments being left empty.
Grants have been available to retailers to take on derelict buildings.	Traffic congestion due to increased numbers of residents.
Young people moving to the city for university and to find work need housing close to the amenities and institutions.	The process of gentrification might mean that working-class people are unable to buy or rent property in the city.
Gentrification has also helped to revive inner-city areas, where the housing offers easy access to work and entertainment in the city centre.	As the image of the city improves, more people are attracted to it.
With better health care in the cities, older people who moved to rural areas when they retired may seek to return to the city for better access to hospitals.	

Now test yourself

1 What encouraged the movement of people into the suburbs of cities?
2 What are the push and pull factors involved in counter-urbanisation?
3 Why are cities experiencing re-urbanisation?

Case study: A city in an advanced country

Case study: Leeds, UK

Location and importance

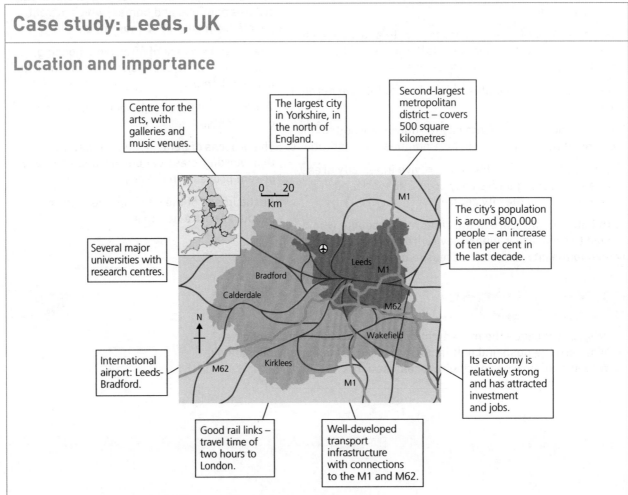

Centre for the arts, with galleries and music venues.

The largest city in Yorkshire, in the north of England.

Second-largest metropolitan district – covers 500 square kilometres

The city's population is around 800,000 people – an increase of ten per cent in the last decade.

Several major universities with research centres.

International airport: Leeds-Bradford.

Good rail links – travel time of two hours to London.

Well-developed transport infrastructure with connections to the M1 and M62.

Its economy is relatively strong and has attracted investment and jobs.

Figure 1 **Location map of Leeds**

Patterns of migration

- Around 17 per cent of the population are from black and ethnic minority communities.
- Residents who are non-UK born tend to settle in the Gipton and Harehills wards of the city.
- There was an influx of migrants, including 'new commonwealth' immigrants from the Caribbean, during the 1950s. The city has a West Indian carnival every year.
- Pakistani and Indian communities were also well represented during the 1950s.
- A large Irish community established themselves in the early nineteenth century. They were spread through the city after the slum clearances, having initially settled in an area called 'the Bank'.

- After a second wave of immigration in the mid-twentieth century, the Irish community numbered over 30,000. They found work in labouring and manufacturing jobs.
- After the Second World War, the city welcomed Polish, Ukrainian and Hungarian refugees and, after the extension of the EU in 2004, new arrivals from Lithuania.
- Around 89 per cent of the population of Leeds were born in the UK.
- In 2013, the Office for National Statistics estimated that there were between 6000 and 9000 new long-term immigrants in the city (net migration was around 1700).

Character of the city

- The city is very diverse. Between 1991 and 2011, the ethnic minority population in Leeds doubled in size. The largest groups were Pakistani and Indian. Some ethnic minorities struggle to gain well-paid work. Pakistani communities show more preference for living in the same area of the city.
- The university influences the character of the city, with 30,000 students and 7000 staff at the University of Leeds alone. The number of young people creates demand for housing stock and they are important for the local economy.
- Leeds has a high proportion of young people; eighteen per cent of the city are aged fifteen and under. Many retail and entertainment businesses cater for the teenage and young adult market. There is a thriving café culture.

Way of life

- **Industry**: the headquarters of ASDA and Danish company ARLA are in Leeds. It has a well-developed digital infrastructure that attracts new business. Many creative industries work from hubs based in old Victorian industrial buildings. M&S opened their first arcade in Leeds in 1884 and grew from there.
- **Sport**: the 2014 Tour de France started in Leeds, with millions of spectators lining the streets of the city. This was part of the Yorkshire tourism agency's bid to raise the profile of the city. There are football and rugby teams, as well as a cricket ground.
- **New developments**: new buildings are being erected along the waterfront and canal. The council plans to build affordable housing to tackle the issue of homelessness. Some businesses change their names during the week to provide different experiences for young people. There has been large-scale redevelopment, though not all developments have full occupancy.
- **Surrounding area**: the city sits close to the Yorkshire Dales and North York Moors National Parks and there are other Areas of Outstanding Natural Beauty (AONB) nearby.

Challenges

- **Social inequality**: the gap between the wealthiest and poorest residents is significant. In Holbeck, over fifteen per cent of residents were on Jobseeker's Allowance and other benefits in 2015, while in Weetwood that figure was just 0.2 per cent. Leeds has been identified as having the third-highest levels of inequality of any city in the UK.
- **Studentification**: this term is used to describe a student community replacing the local community. This is true of South Headingley and Hyde Park. It creates a d emographic imbalance and a rise in property prices. There are often more pubs and takeaway restaurants in those areas, and crime tends to be high, particularly from antisocial behaviour. Pride in the community and the appearance of housing also tends to be reduced.
- **Transport issues**: there is a demand for a better mass transportation system; fares are rising and there is a need for a well-developed tram system to reach more areas of the city.
- **Loss of local businesses**: large retail developments such as Eastgate and Trinity have taken up local investment at the expense of local businesses, leading to the loss of independent, local businesses.

Sustainable initiatives

Plans for the South Bank

- Infrastructure and investment, including a new HS2 railway station
- Supports retail, leisure and financial services
- Cultural centre at the old Tetley brewery will support contemporary art
- Educational improvements linked to Leeds City College, particularly for vocational courses
- A new 3.5-hectare park and open space along the waterfront

- Over 300,000 square metres of development land available
- Creation of Holbeck Urban Village to improve the physical and social environment
- New pedestrian and cycle bridges
- Clarence Dock will become Leeds Dock and contain entertainment, restaurants and retail developments
- Water taxis and shuttle buses will connect the area, reducing carbon dioxide emissions.

Exam practice

With reference to your chosen city case study, examine how ways of life can vary within one AC city. [8]

ONLINE

Case study: A city in an emerging developing country

Case study: Rosario, Argentina

Location and importance

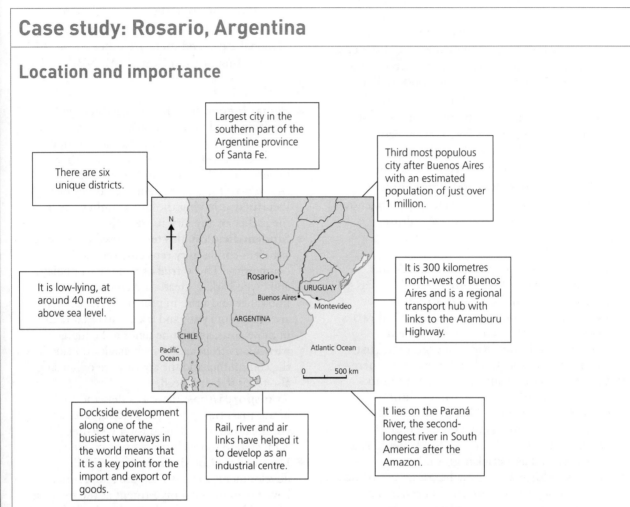

There are six unique districts.

Largest city in the southern part of the Argentine province of Santa Fe.

Third most populous city after Buenos Aires with an estimated population of just over 1 million.

It is low-lying, at around 40 metres above sea level.

It is 300 kilometres north-west of Buenos Aires and is a regional transport hub with links to the Aramburu Highway.

Dockside development along one of the busiest waterways in the world means that it is a key point for the import and export of goods.

Rail, river and air links have helped it to develop as an industrial centre.

It lies on the Paraná River, the second-longest river in South America after the Amazon.

Figure 2 Location map for Rosario, Argentina

Patterns of migration

- The culture of Rosario has been enriched by the fact that it has attracted people from across Argentina.
- Arrivals included the Spanish in the sixteenth century and migrants from across Europe in the nineteenth century.
- More recently, there has been an influx of migrants from other countries within South America and from as far away as China and Taiwan.
- Italians have formed a significant number of immigrants to Rosario over the last century, which has influenced the city's culture, particularly the food and architecture.
- The city has a young demographic with a relatively high birth rate, though there are signs of an ageing population across the city.

Way of life

- The city has been described as the socialist hub of the country. It has close links with trade unions due to its industrial heritage.
- Residents and visitors have good opportunities for shopping in El Centro, the central mall.
- Argentina has a close cultural connection to meat, with one of the highest meat consumptions per capita in the world. The country is the third-largest exporter of meat products. Cattle graze the large grassland areas and the gauchos (cowboys) are still important. The *asado*, or grill, is an important feature of many restaurants.
- The Argentine flag comes from Rosario and was raised in the city for the first time. The National Flag Memorial is located on the bank of the Paraná.

Challenges

- **Unemployment**: there were riots in the city as recently as 2001, with high unemployment rates and economic problems leading people to loot supermarkets. As the employment situation improves, levels of crime drop.
- **Crime**: slum districts, known as *villas miserias*, are beset by high levels of poverty and crime. There are violent drug wars in some districts, and criminals have allegedly infiltrated the police and football teams to take control.
- **Social inequality and divisions**: the slums house 100,000 people and occupy ten per cent of the space in the city. The city's infrastructure cannot keep pace with the new arrivals.

Sustainable initiatives: Pro Huerta

- The city has gained a reputation as a world leader in urban forestry, reducing climate change in the city. The programme is called *Pro Huerta* (literally, Pro Garden).
- Trees are planted among all new housing developments to reduce the temperature of the area.
- Flood risk zones and semi-urban land have been planted with food crops, which improves the diet of some low-income families as well as providing additional income from the sale of their produce.
- Community groups have been given tools, seeds and vacant land to cultivate crops.
- Producing food locally reduces the need to transport crops into the city, thus reducing carbon dioxide levels.
- Local hotels and restaurants reduce their food miles and promote the value of buying locally by supporting urban farms.
- There are over 800 community gardens in Rosario, supporting over 40,000 people.
- Rosario won a UN Habitat Award for its work in this area.
- Most of the gardeners are women, who earn an income and gain a skill.

> **Exam tip**
>
> You only need to know one sustainable initiative, in detail, for each city.

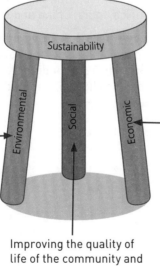

Buildings that are energy efficient, use of renewable energy and conservation of habitats.

Employment with fair wages and which is close to home to avoid lengthy travel times.

Improving the quality of life of the community and the ways in which people in the area interact.

Figure 3 What makes a project 'sustainable'?

> **Exam tip**
>
> Students often refer to many examples when a question clearly specifies **one**. This is such a common mistake that it is normally marked in bold to draw your attention to it.

Exam practice

With reference to your chosen city case study, examine the challenges faced within **one** LIDC or EDC city. [8]

 ONLINE

11 Why are some countries richer than others?

The definition of development and how countries are classified

The definition of development

The term **development** can be used to describe the progress of a country as it becomes more economically and technologically advanced. It can also be applied to improvements in people's quality of life – educational opportunities, increased incomes, human rights and healthy living conditions.

> **Development**: the progress of a country in terms of economic growth, the use of technology and human welfare.

How countries can be classified

Countries have been classified by global organisations such as the World Bank (WB), the United Nations Development Programme (UNDP) and the International Monetary Fund (IMF). A range of economic and social indicators is used to split the world up into groups that are broadly similar.

The OCR GCSE and A level specifications use the IMF classification:

> **Advanced countries (ACs)**: well-developed financial markets; diversified economic structure with rapidly growing service sector; examples include the UK, USA, Japan and Australia.
>
> **Emerging and developing countries (EDCs)**: do not share all the characteristics required to be an AC but are not eligible for Poverty Reduction and Growth Trust funding; examples include South Africa, India, China and Brazil.
>
> **Low-income developing countries (LIDCs)**: countries eligible for Poverty Reduction and Growth Trust funding from the IMF; examples include Nigeria, Bangladesh and Afghanistan.

Global distribution of ACs, EDCs and LIDCs

Figure 1 is a development map of the world based on the IMF's definitions. Notice that most LIDCs are in Africa, with a few in the Middle East, Asia and South America.

Revision activity

Draw a summary table to define each of the IMF development categories. Use Figure 1 to include a selection of countries for each category.

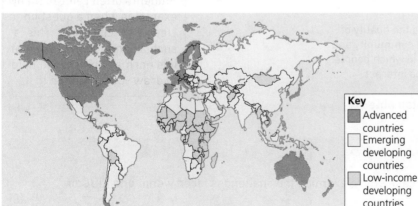

Key
- Advanced countries
- Emerging developing countries
- Low-income developing countries

Figure 1 International Monetary Fund (IMF) development classification

How development is measured

Measures of development

There are several economic and social measures of development.

- **Economic measures**: these are to do with money and include gross domestic product (GDP), gross national income (GNI), and various monetary measures of poverty and standard of living.
- **Social measures**: these are to do with people and include infant mortality, life expectancy, access to doctors, and educational attendance and achievement.

While there are significant similarities between the global patterns produced, there are subtle variations, and some indicators tend to be more reliable than others.

It is important to remember that measures are averaged for a whole country. There will often be significant inequalities of wealth and social development *within* a country, particularly between major cities and remote rural areas. In fact, inequality is a good measure of the lack of development of a country.

Figure 2 Measures of development: variations and limitations

Measure of development	Global variations	Limitations
Gross national income (GNI) per capita	There are huge global variations, for example Norway ($93,820) and Somalia ($150); most of the poorest countries are in Africa.	These average figures can be misleading – a few very wealthy people in a country can distort the figures. In poorer countries many people work in farming or in the informal sector, where their income is not taken into account by official GNI records. Data about income is sensitive and people may not always be honest.
Birth rate (number of live births per 1000 population)	High birth rates are generally associated with poorer countries where child survival rates are low (due to poor health care, lack of safe water, poor diets and sanitation). Large families ensure a decent income for the family and provide support for ageing parents. The highest birth rates exceed 40 per 1000 (for example, 44 per 1000 in Afghanistan), with the lowest rates being about 10 per 1000 (for example, 11 per 1000 in Finland).	Birth rates are quite a good measure of economic and social development. However, some countries have a low birth rate even though most people are relatively poor (for example, Cuba with 10 per 1000). This is due to political decisions to focus investment in health care over other sectors. Birth control policies can also distort this as a measure of overall development (for example, China with 12 per 1000).
Infant mortality (number of deaths of children aged less than one year of age per 1000 population)	Figures vary enormously, with the highest values in African countries (for example, Angola has 96 per 1000) and the lowest values in ACs (for example, Germany with 3 per 1000).	Recognised as a good measure of development as it reflects the levels of health care and service provision in a country. Not all deaths of children are reported in the poorest countries, especially in remote areas. The true rates may be even higher.

Figure 2 Measures of development: variations and limitations

Measure of development	Global variations	Limitations
Life expectancy (average number of years a person can be expected to live at birth)	In ACs, life expectancy can be over 80 years; in EDCs, life expectancy is between 65 and 75 years; in LIDCs, life expectancy is typically in the 50s (for example, in Nigeria it is 53).	This is generally a good measure as it reflects health care and service provision. Data is not always reliable, especially in LIDCs, and it can be slightly misleading in countries with very high rates of infant mortality – people surviving infancy may live longer than expected thereafter.
Literacy rates (percentage of people with basic reading and writing skills)	Most ACs have literacy rates of 99%; in LIDCs, the figure can be below 50% (for example, Afghanistan has a rate of 38%).	Another good indicator of development though it can be hard to measure, especially in LIDCs, due to the lack of monitoring. War zones and squatter settlements are difficult areas in which to measure literacy rates.
Human Development Index (composite measure using data on income, life expectancy and education to calculate an index from 0–1)	Highest HDI values are found in ACs (Norway 0.944); the lowest are found in the African LIDCs (Niger 0.348).	Developed by the UN, this is the most commonly used measure of development.

The Human Development Index

REVISED

Figure 3 shows global development according to the widely used Human Development Index. Figure 4 is a topological map showing $GNI per capita. Notice that the area of each country is proportional to the country's GNI. Such maps are visually very powerful and show disparities very clearly.

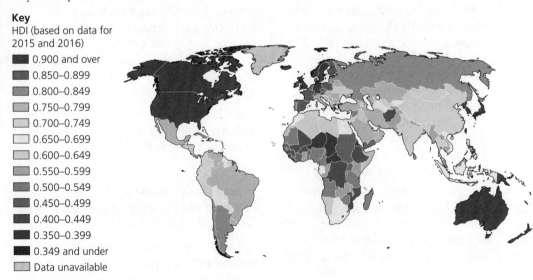

Key

HDI (based on data for 2015 and 2016)

- 0.900 and over
- 0.850–0.899
- 0.800–0.849
- 0.750–0.799
- 0.700–0.749
- 0.650–0.699
- 0.600–0.649
- 0.550–0.599
- 0.500–0.549
- 0.450–0.499
- 0.400–0.449
- 0.350–0.399
- 0.349 and under
- Data unavailable

Figure 3 Global development according to the Human Development Index

GNI per capita

Figure 4 Topological map showing $GNI per capita

Figure 4 is taken from www.worldmapper.org, which uses a consistent colour scheme in all of the maps they produce. They divide the world into twelve separate regions (North America, South America, Western Europe, Eastern Europe, Northern Africa, Central Africa, Southeastern Africa, Middle East, Southern Asia, Eastern Asia, Japan and Asia Pacific) and use twelve colours that vary in shades to identify territories within regions. The twelve regions are coloured in order from poorest to richest by the Human Development Index, with shades of dark red to show the poorest regions, going through the rainbow spectrum of orange, yellow, green and blue, with a shade of violet for the most well-off regions.

Now test yourself

1 What are the limitations of using gross national income (GNI) as a measure of development?
2 Why is life expectancy a good measure of development?
3 Is birth rate a good measure of development?
4 The Human Development Index is a 'composite' index. What does this mean and why does it make the HDI one of the most widely used measures of development?

Exam practice

1 What is meant by the term 'development'? [2]
2 Study Figure 1 on page 88. Describe the pattern of Low-income developing countries (LIDCs). [4]
3 Examine how economic and social measures can be used to illustrate the consequences of uneven development. [6]

Exam tip

In the exam it is extremely likely that you will be given a map using social and/or economic measures to show global development. Make full use of the map when answering the question. Refer to global regions and specific countries, using the key to give specific values.

Factors influencing uneven development

Physical causes

The physical geography of a country or region can create challenges for economic development and put it at disadvantage:

- **Weather and climate**: heavy rainfall, droughts, extreme heat or cold and vulnerability to tropical cyclones hamper economic development. Vast parts of central and western Africa experience limited and unreliable rainfall. The Philippines and the Caribbean are frequently ravaged by tropical storms. In 2016, over 1000 people were killed by Hurricane Matthew in Haiti, just six years after 230,000 people were killed by a powerful earthquake.
- **Relief**: mountainous countries and regions, for example Nepal, tend to be remote and have a poor infrastructure. They are also subject to extreme weather conditions.
- **Landlocked countries**: countries without a coastline lack the benefits of sea trade, which historically has led to the development of most of the world's most-developed nations. A coastline acts as an international border providing huge opportunities for trading with other nations. Eight out of the fifteen lowest-ranking countries according to the HDI are landlocked (Figure 5).
- **Tropical environment**: tropical environments (hot and wet) are prone to pests and diseases, which can spread rapidly. Malaria, spread by mosquitoes, and water-borne diseases such as cholera, can devastate communities and reduce people's ability to work.
- **Water shortages**: water is essential for life and for development. There are serious shortages of water in some parts of the world, for example in parts of Africa and the Middle East.

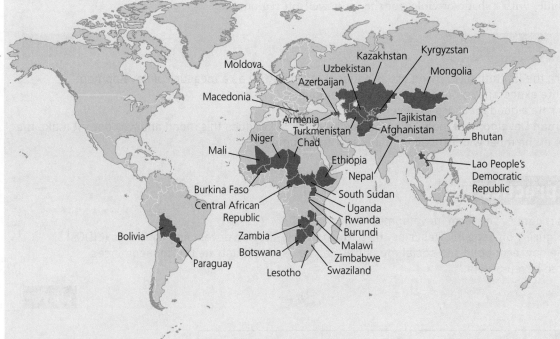

Figure 5 Landlocked developing countries (according to the UN)

1 How can weather and climate affect economic development?
2 Study Figure 5.
   (a) Describe the distribution of landlocked developing countries.
   (b) Why does the lack of a coastline hinder economic development?

Human causes

Several human factors affect development, including political stability, technology, health care and cultural traditions. Three of the most important factors are poverty, trade and history.

Poverty

- The lack of money in a household, community or country slows development.
- It prevents improvements to living conditions, infrastructure and sanitation, education and training.
- Without the basics, developments in agriculture and industry will be extremely slow and an economy will simply fail to take off. Figure 6 illustrates the cycle of poverty.

Trade

- Trade between nations involves the import and export of goods and services.
- ACs: the vast majority of the world's trade involves the richer countries of Europe, Asia and North America. Most of the world's powerful **transnational companies (TNCs)** are based here.
- LIDCs: poorer countries have limited access to the markets. They have traditionally traded relatively low-value raw materials, such as agricultural products or minerals, rather than higher-value processed goods. The value of these raw materials (commodities) can fluctuate wildly, causing great uncertainty and instability as countries strive to become developed (Figure 7).

History

- Many ACs have experienced a long history of development based on agricultural and industrial growth and international trading.
- This has allowed them to become highly developed and relatively wealthy.
- Rapid industrialisation has taken place in EDCs such as China, Malaysia and South Korea in recent decades.
- The LIDCs have not experienced significant economic growth yet. Many LIDCs were colonised by powerful trading nations such as the UK and France. It was during this colonial era that global development became uneven. The colonial countries face huge challenges, including poor infrastructure, lack of administrative experience and political instability.

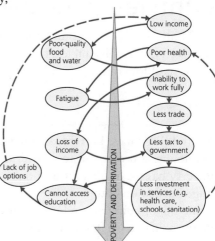

Figure 6 The cycle of poverty

> **Transnational companies (TNCs)**: global organisations that operate around the world but usually have their headquarters in ACs.

> **Revision activity**
>
> Create a spider diagram to summarise the main physical and human causes of uneven global development.

> **Now test yourself**
>
> 1 With reference to Figure 6, explain why 'poverty leads to poverty'?
> 2 How has colonialism hindered economic development in many LIDCs?
>
> TESTED

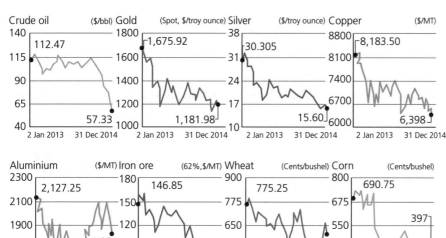
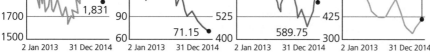

Figure 7 Recent fluctuations in commodity prices

11 Why are some countries richer than others?

Factors that make it hard for countries to break out of poverty

Over 3 billion people – just under half the world's population – suffer from poverty. They live on less than $2.50 a day. There are several reasons why it is hard for countries and individuals to break out of poverty.

Debt

REVISED

- Many LIDCs have borrowed money at high rates of interest to pay for development projects or to recover from natural disasters.
- Poverty has meant that most of these debts are still unpaid, placing a huge burden on countries.
- Some donor countries and organisations have cancelled or reduced debts to enable LIDCs to develop.

Trade

REVISED

- Global trade favours richer countries that have formed powerful trading groups (blocs). Most value adding takes place in these countries.
- TNCs tend to be based in ACs, exploiting the LIDCs for their raw materials and relatively cheap labour force.
- For LIDCs to develop, they need to have a better balance of trade, with valuable exports being traded with other countries. Trade also needs to be fair, so that producers get a fair income for what they produce.

Political unrest

REVISED

- Many LIDCs have unstable and corrupt governments with limited levels of democracy.
- Businesses are reluctant to get involved in unstable countries, so investment will be limited.
- Corruption results in a lack of internal investment in services, health care and education.
- Some LIDCs are experiencing civil wars, tribal disputes and terrorism.

Now test yourself

Explain why political unrest makes it hard for some countries to break out of poverty.

TESTED

Exam practice

1 Study Figure 7.
  (a) Describe the pattern of commodity prices shown by the graphs. [4]
  (b) Suggest why a reliance on the export of commodities hinders economic development. [4]
2 To what extent is uneven development the result of physical factors? [6]
3 Evaluate the factors that make it hard for countries to break away from poverty. [6]

ONLINE

Exam tip

When evaluating Exam practice question 3, you need to consider and weigh up the good *and* the bad, the advantages *and* disadvantages. Your answer needs to be balanced. You should be prepared to express and justify your own opinion; for example, do you think that one factor is more important than all the others?

Now test yourself and exam practice answers at **www.hoddereducation.co.uk/myrevisionnotes**

12 Are LIDCs likely to stay poor?

This chapter uses Ethiopia as a case study for the economic development of an LIDC.

Economic development in Ethiopia
Where is Ethiopia?

REVISED

Ethiopia is located in the centre-east of Africa and is bordered by six other countries (Figure 1). Some fact about Ethiopia:

- It is the continent's tenth-largest country by area and second most populous after Nigeria.
- Ethiopia's landscape varies from the densely vegetated Western Highlands to arid desert in the Eastern Lowlands.
- The country has suffered from periodic drought and famine for decades.
- Ethiopia is Africa's oldest independent country (it remained independent throughout the colonial period) and was a founder member of the UN.

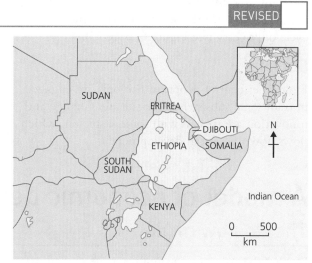

Figure 1 The location of Ethiopia

How has Ethiopia's economy developed?

REVISED

Ethiopia is an LIDC and, with an HDI of just 0.435, it is one of the world's poorest countries (Figure 2). With a GNI of just $505 per capita (2015), average incomes are significantly lower than the world average of $10,858 per capita.

Figure 3 shows the very slow growth in Ethiopia's wealth since 2004. Ethiopia is also significantly less wealthy than Sub-Saharan Africa and other LIDCs across the world.

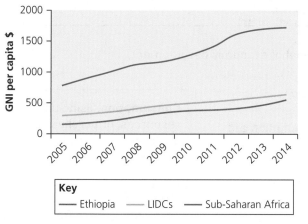

Figure 3 Ethiopia's wealth over time

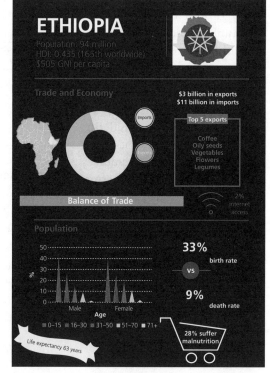

Figure 2 Ethiopia in numbers

Revision activity

Use Figures 2 and 3 to compile a brief fact file describing Ethiopia's economic development.

Ethiopia's economic growth rate

There are several reasons why Ethiopia's growth rate is so slow and why it remains one of the world's poorest countries.

Population

Ethiopia has a very large and growing (2.6 per cent per year) population of over 94 million people. As health care gradually improves, death rates decline, life expectancy increases and the population expands. This means that more people will be using the same amount of resources, such as food, water, education and health care. Ethiopia is stuck in a cycle of poverty (see Figure 6, page 93).

Society

Eighty per cent of people in Ethiopia are engaged in traditional subsistence agriculture. Change and innovation are treated with some suspicion, which slows down potential development.

Technology

There is very limited development of technology (industry, infrastructure, communications) in Ethiopia due to limited investment and low incomes.

Politics

There has been a history of civil war and political unrest in the country, which has had a damaging effect on development. In the period 1999–2000 there was a border war with Eritrea. There has been a relatively stable government since 2012, and economic development has taken place.

A model of economic development

The Rostow model

In 1960 the American economist Walt Rostow created a model to show how countries progress through different stages of development (Figure 4). Notice that the model consists of a number of stages or steps that ultimately result in the development of an advanced and wealthy economy.

While Rostow's model offers a useful framework for considering economic development, it assumes equal access to resources and a degree of rigidity in terms of progression. In reality, some countries have skipped stages to progress to an advanced economy. Rostow's model applied to the reality:

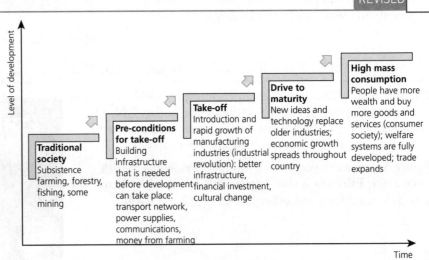

Figure 4 Rostow's model of economic development

- Most LIDCs are in the first two stages of the model.
- ACs are in stages 4 and 5, experiencing high levels of **consumption** and with the majority of people employed in the tertiary and quaternary sectors.
- Ethiopia is probably in stage 2 of the model. While it is still a largely traditional society dominated by agriculture, there have been advances in technology and improvements in education and health care.

Consumption: using up resources, or purchasing goods and produce.

Exam tip

Take time to learn Rostow's model so that you can draw a simple sketch. This model is central to a lot of the work on Ethiopia, so you do need to know and understand it.

Now test yourself

1 Draw a simplified version of Rostow's model of economic development.
2 Can Rostow's model be successfully applied to Ethiopia?

Millennium Development Goals

In 2000, the United Nations Millennium Summit committed to a set of targets to reduce poverty by 2015. Known as the **Millennium Development Goals** (Figure 5), these targets have led to significant improvements in the quality of life for many people in LIDCs. The greatest progress has been in reducing child mortality and increasing school enrolment.

> **Millennium Development Goals**: a set of targets to reduce poverty by 2015.

Have the Millennium Development Goals been met in Ethiopia?

REVISED

In Ethiopia, massive government spending has led to improvements in health and education. The following goals have been met:

- **MDG 2, Primary education:** primary school enrolment increased from 50 per cent in 1990 to 96 per cent in 2015. Literacy rates are still low (36 per cent), indicating that the quality of education provision needs to improve.
- **MDG 4, Child mortality:** infant mortality has more than halved from 97 per 1000 births in 1990 to 45 per 1000 births in 2015. Around 65 per cent of children receive vaccinations. Improvements still need to be made in some rural areas.
- **MDG 5, Maternal health:** improved health care has reduced mortality rates by 23 per cent, and 55 per cent of women now have access to contraception.

The following goals have not been met:

- **MDG 1, Poverty and hunger:** despite improvements, nearly 30 per cent still live in poverty and 40 per cent of children are malnourished.
- **MDG 3, Gender equality:** despite improvements in political representation and education, unemployment is high and many women still carry out traditional tasks (for example, water carrying).
- **MDG 6, Combat disease:** malaria remains a major threat despite 100 per cent of the population being able to access mosquito nets. Access to doctors is poor, with one doctor for every 3333 people. Access to safe water has increased to 69 per cent but waterborne disease is still a major problem due to insufficient sanitation systems.
- **MDG 7, Environmental sustainability:** soil erosion and desertification are widespread, resulting from drought and over-farming.
- **MDG 8, Global partnerships:** national debt has decreased but Ethiopia still relies on a significant amount of foreign aid.

Figure 5 The Millennium Development Goals

Revision activity

Construct a table with eight boxes to represent the Millennium Development Goals. Describe each goal and the extent to which Ethiopia has met its aims.

Wider political, social and environmental factors affecting Ethiopia's development

Factors affecting Ethiopia's development

Figure 6 Political, social and environmental factors that have affected Ethiopia's development

Political	Social	Environmental
Significant political unrest – including coups and civil wars – over the last 50 years has led to inequality in land ownership, poor agricultural productivity and huge social problems.	Many women, particularly in rural areas, maintain traditional roles (such as fetching water).	Ethiopia experiences unpredictable and unreliable rainfall, affecting farming and water supply.
Increasingly productive farmland has been purchased by wealthy foreigners (especially from Saudi Arabia) for commercial (export) crops.	Equality between men and women is still patchy in education and employment.	Droughts are a constant threat, particularly in the Eastern Lowlands where the average annual rainfall is below 300 mm. A severe drought and subsequent famine killed over one million people in 1984–5.
The current Ethiopian government has addressed the Millennium Development Goals by investing in education, health and agriculture.	Very few women attend secondary schools or follow careers, keeping their level of education low.	Water shortages have led to overgrazing and soil erosion, with desertification being an issue on marginal land.
The USA provides considerable financial support to the agricultural sector.	Rural areas are somewhat backward compared to urban areas. Much of the rural economy depends on traditional subsistence farming.	Lowland regions suffer from mosquitoes and tsetse flies, spreading diseases such as malaria and sleeping sickness.
Recent Chinese investment has led to improvements in infrastructure, such as roads.	Some people are unwilling to access contraception despite its widespread availability – this has an impact on birth rates, infant mortality and levels of poverty.	The Western Highlands experience wet and warm conditions, enabling crops to grow well. However, some areas are mountainous with steep slopes and thin soils.

Revision activity

Use a highlighter to identify **three** political, social and environmental factors that have affected Ethiopia's economic development. Choose factors that you think you will be able to learn and write about in an exam question.

Exam tip

Exam practice questions 3 and 4 require you to express your opinion and to enter into a discussion. Remember to weigh up both sides of the argument and try to present a balanced and objective assessment. Be prepared to justify your point of view and refer to specific facts and figures relating to your case study (Ethiopia).

Exam practice

1 Examine how population has affected Ethiopia's economic development. [4]
2 (a) Outline the main characteristics of Rostow's model of economic development. [4]
  (b) To what extent can Rostow's model help us to understand Ethiopia's path to economic development? [6]
3 'Ethiopia has made significant progress in meeting the Millennium Development Goals.' Do you agree with this statement? [6]
4 Examine the extent to which environmental factors have affected Ethiopia's economic development. [6]

ONLINE

International trade

Ethiopia has a trade deficit: in 2015, it exported goods to the value of $5.44 billion but imported goods worth $17.6 billion. In order to move through the stages in Rostow's model, this trading deficit needs to be reduced. This will enable the country to spend more money on improvements to infrastructure, education and health care.

Ethiopia's exports

Currently Ethiopia's exports are dominated by agricultural produce, particularly coffee, food and flowers (see below). Most of the crops are grown in the wetter and more productive Ethiopian highlands.

Dependence on commodities means that the economy is vulnerable to factors affecting production, such as weather and climate, as well as global economics that affect world prices. Issues with ground transport and storage will also affect the quantity and quality of food exports. Much of what is sold abroad is of low value, with only limited 'value-adding' processing taking place in Ethiopia (see also Ethiopia's future economic growth, page 100).

Ethiopia's exports, 2015

Total: $5.44 billion

- Coffee 17%
- Refined petroleum 11%
- Cut flowers 13%
- Gold 11%
- Other vegetables 10%
- Other oily seeds 8.7%
- Dried legumes 4.3%
- Bovine 3.2%
- Sheep and goat meat 1.9%
- Gas turbines 1.9%
- Sheep and goats 1.5%
- Other animals 1.1%
- Tanned sheep hides 1.1%
- Knit t-shirts
- Aircraft parts
- Other footwear

> **Revision activity**
>
> What were the top five exports in Ethiopia in 2015? Apart from agricultural products, what are Ethiopia's other main exports?

Ethiopia's imports

Ethiopia's top five imports are petroleum, trucks, fertilisers, construction and wheat. These imports – petroleum to fuel industry and transport, trucks to improve transport, fertilisers to improve agricultural production and construction equipment for building – indicate that Ethiopia is striving for economic development.

With the exception of wheat, these are all processed, high-value products, which explains Ethiopia's trade deficit. It is perhaps ironic that Ethiopia exports agricultural products but still has to import wheat to feed its people.

Ethiopia's trading partners

Ethiopia has strong trading relations with several countries, including Somalia, China, Kuwait, India and Switzerland. These global links are evidence of a stable government and economy, and will support Ethiopia's future economic development.

Figure 7 Ethiopia's main trading partners, 2015

Main nations that Ethiopia exports to	Percentage of total exports ($5.44 billion in 2015)	Main nations that Ethiopia imports from	Percentage of total imports ($17.6 billion in 2015)
Somalia	12%	China	33%
Kuwait	12%	Kuwait	8%
Switzerland	11%	India	7%
Netherlands	11%	USA	5%
China	7%	Japan	4%

Ethiopia's future economic growth

In order for future economic growth to take place, Ethiopia needs to become less dependent on the export of low-value commodities. It needs to develop its processing and manufacturing sectors by encouraging investment from TNCs.

Ethiopia also needs to develop its tertiary sector (health care, education, support services, tourism) if it is to move into stages 3 and 4 of Rostow's model. There are many attractions for tourists, and the potential for development in this sector is huge.

Now test yourself

1 What are Ethiopia's major exports and imports?
2 What are the potential problems with an overdependence on the export of agricultural products?
3 Why is it important for Ethiopia to have strong trading links with foreign countries?

The role of TNCs

TNC investment in Ethiopia

TNCs are huge global organisations that operate around the world but usually have their headquarters in rich countries (ACs). There are several TNCs operating in Ethiopia (Figure 8). Most are involved in manufacturing, investing money, materials and expertise that would not be readily available in the country.

Figure 8 TNC investment in Ethiopia

Company	What do they do in Ethiopia?
Hilton Hotels	Leisure and recreation services, hotel creation
Siemens	Manufacturing of telecommunications, electrical items and medical technology
General Electric (GE)	Aviation manufacturing, delivering rail links
Afriflora	Flower growing – the world's largest producer of fair-trade roses
Dow Chemicals	Manufacturing chemicals, plastics and agricultural products
H&M	Textiles manufacturing, university education in textiles

Advantages and disadvantages of TNC investment

Figure 9 Advantages and disadvantages of TNC investment

Advantages of TNCs	Disadvantages of TNCs
Large companies provide employment and training	TNCs can exploit the low-wage economy and avoid paying local taxes
Modern technology is introduced	Working conditions may be poor, with fewer rules and regulations than exist in richer countries
Companies often invest in the local area, improving services (for example roads and electricity) and social amenities	Environmental damage may be caused
Local companies may benefit by supplying the TNCs	Higher-paid management jobs are often held by foreign nationals
TNCs have many international business links, helping industry to thrive	Most of the profit goes abroad rather than benefiting the host country
The government benefits from export taxes, providing money that can be spent on improving education, health care and services	Incentives used to attract TNCs could have been spent supporting Ethiopian companies

Revision activity

Create a spider diagram to summarise some of the advantages and disadvantages of TNCs.

Aid and debt relief

Ethiopia has benefited significantly from international aid and debt relief.

Aid

Ethiopia has benefited from short-term emergency aid – most notably during the severe famine of the 1980s – and longer-term developmental aid. Aid comes from:

- individual countries, such as the UK, USA and Russia (unilateral aid)
- organisations such as the World Bank
- non-governmental organisations (NGOs), such as the charities Oxfam, Farm Africa and Mission Aviation Fellowship.

Aid has been successful in creating real improvements in roads, schools and water supply, benefiting the poorest people for whom the aid is intended. The fact that there appears to be little corruption has encouraged countries and organisations to continue to donate.

Through its Goat Aid programme (Figure 10), Oxfam encourages young girls to raise goats. It enables them to benefit from an improved diet, invest in educational opportunities and raise their status. In supporting equality and reducing birth rates, this sustainable initiative supports long-term economic development.

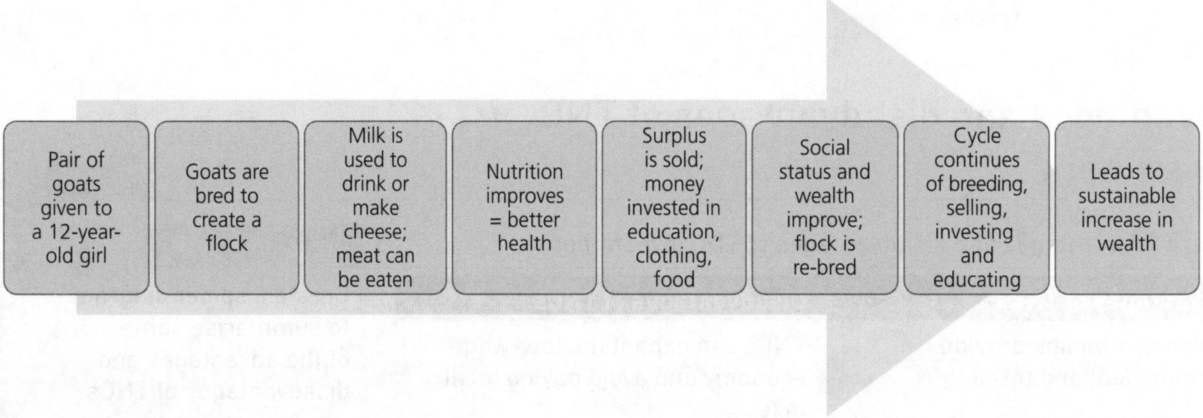

| Pair of goats given to a 12-year-old girl | Goats are bred to create a flock | Milk is used to drink or make cheese; meat can be eaten | Nutrition improves = better health | Surplus is sold; money invested in education, clothing, food | Social status and wealth improve; flock is re-bred | Cycle continues of breeding, selling, investing and educating | Leads to sustainable increase in wealth |

Figure 10 Oxfam's Goat Aid programme

Debt

In 2001 Ethiopia qualified for the World Bank/IMF Highly Indebted Poor Countries debt relief programme. In total, debts have been reduced by about $2 billion from wealthy countries and organisations such as the World Bank. The former countries of the Soviet Union, including Russia, have cancelled $5 billion of debt, reducing Ethiopia's debt by a half.

Ethiopia's current debt is about $12–14 billion, representing about 23 per cent of Ethiopia's GDP. This level of debt is not considered to represent a major problem in the long term by the World Bank and the IMF.

Much of the government's borrowing is being spent on transport and communication infrastructure, electricity production and agriculture. These sectors have a high rate of return on investment and contribute to economic growth. This means that the population will continue to benefit from these investments in the future.

Ethiopia's public external debt (about $5.6 billion) is owed by public enterprises that have invested significantly in industrial development projects:

- Ethiopian Electric Power Corporation (EEPCO) has constructed several dams to supply electricity and manage water supplies.
- Ethiopian Sugar Corporation is building ten sugar refineries.
- Ethiopian Railways Corporation aims to build 5000 kilometres of railway to improve the nation's network.

Now test yourself

1 Why do countries such as Ethiopia get into debt?
2 How can debt relief help Ethiopia to develop its economy?
3 Why is borrowing money not necessarily a bad thing?

Exam practice

1 Examine the importance of international trade in affecting economic development in Ethiopia. [6]
2 (a) What are transnational companies? [2]
  (b) Why are they attracted to locate factories in Ethiopia? [4]
3 Assess the advantages and disadvantages of **either** aid **or** debt relief for promoting economic development in Ethiopia. [6]

Exam tip

In Exam practice question 3, you must write a balanced answer referring to both advantages *and* disadvantages. Make sure that you only write about aid *or* debt relief – not both.

Development strategies

It is possible to identify two broad types of development strategy: **bottom-up strategies** and **top-down strategies**.

> **Bottom-up strategies**: small-scale initiatives, usually led by the local community.
>
> **Top-down strategies**: large-scale initiatives, usually controlled by the government.

Bottom-up strategies

REVISED

Bottom-up strategies are often low-cost schemes funded by charities. They are often small-scale, highly localised improvements that are initiated and led by local communities. There are several examples of bottom-up strategies in Ethiopia:

- Farm Africa has worked with rural communities to breed goats and chickens, and to construct beehives. Once established, people are encouraged to donate animals to other communities nearby, thereby spreading development.
- Abyssinian Flight Services flies tools to local farming communities.

These highly focused small-scale projects usually achieve a high level of success and can spread development effectively throughout deprived areas.

Top-down strategies

REVISED

Top-down strategies are driven by national governments and often involve expensive, high-impact projects. They usually involve development plans administered and funded by the Ethiopian government.

- Currently, about 60 per cent of the country's national income is spent on development projects to improve education, health care and infrastructure.
- The Growth and Transformation Plan has invested money on infrastructure ($3.6 billion has been spent upgrading rural mud roads to asphalt roads) to enable industry to develop.
- Controversial hydroelectric power plants have been constructed on the Omo River, partly funded by the Chinese government. While promising cheap electricity to promote industrial development, many people have been displaced and vast areas of farmland flooded. Lake Turkana, an important archaeological site, may dry up and water pollution may result.

Now test yourself

TESTED

Describe one bottom-up strategy and one top-down strategy in Ethiopia.

Advantages and disadvantages of development strategies

Figure 11 Advantages and disadvantages of different development strategies

	Advantages	Disadvantages
Bottom-up	Led and managed by local communities having identified a particular need – much more likely to be accepted by people and to have a successful outcome. Relatively low cost, with initial costs often borne by charities. Likely to have minimal impact on the environment. Maintenance will be low cost and will probably be carried out by local people.	Small scale and highly localised means that there are few economies of scale that would be achieved with a larger-scale project. Projects may not be integrated and may even conflict with other schemes in the area. Possible resentment and competition between local communities to receive aid.
Top-down	May involve large amounts of money and technical expertise, which can benefit large numbers of people. Schemes can tackle major issues such as energy, transport and education. Schemes can be integrated – for example, dams and reservoirs can generate electricity and provide water for drinking and irrigation. Prestigious schemes show the government's intent to develop, which can be appealing to foreign investors.	Local people may not be consulted and may even be displaced by new developments. Projects may focus on benefiting the wealthy – large landowners, industries and urban areas. Projects are very costly and often require expensive maintenance. They may not be sustainable. Environmental damage may occur, resulting in soil erosion and desertification.

Revision activity

Create a table to outline the advantages and disadvantages of the Omo River HEP top-down development strategy.

Exam practice

1 Outline the difference between bottom-up and top-down development strategies. [4]
2 Compare the advantages and disadvantages between **one** bottom-up and **one** top-down development strategy in Ethiopia. [6]

ONLINE

Exam tip

The specification (and Exam practice question 2) requires you to be able to compare the advantages and disadvantages of one top-down **and** one bottom-up strategy in an LIDC (Ethiopia). You must make sure that you learn some specific details.

The characteristics of the UK

Relief

REVISED

Find a map that shows the physical geography or **relief** of the British Isles. This could be in an atlas or a textbook. The map should show hills and mountains, lowlands and rivers. Notice the following characteristics:

- Most mountains are located in the north and west of the UK, especially in Wales and Scotland. They are spectacularly beautiful and popular with tourists, but mountainous regions can also be bleak and hostile with few roads and settlements.
- Much of the south and east of the UK is relatively flat with a few hilly areas. With a more moderate climate, this gently rolling landscape is well suited to farming and the development of settlements, roads and railways. It is more densely populated than the UK's mountainous regions.
- There are a lot of rivers in the UK, most of which flow from the hills or mountains down to the sea. The longest river in the UK is the River Severn (354 km), which is just longer than the River Thames (346 km). The largest lake in the UK is Loch Neagh in Northern Ireland, which has an area of 383 square kilometres (roughly equivalent to 38,300 football pitches).

> **Relief**: differences between the high areas and low areas of land.

> **Now test yourself**
>
> Use an atlas map showing the physical geography of the British Isles to answer the following questions.
> 1 Describe the location of the Cambrian Mountains, the North West Highlands and the North Downs.
> 2 Describe the course of the River Thames.
>
> TESTED

Rainfall

REVISED

Figure 1 shows the distribution of precipitation in the British Isles. Notice that the highest rainfall totals are in the north and west, particularly over the mountains. Here the average annual rainfall totals exceed 2000 mm. The lowest precipitation totals are in the south and the east. Here the driest areas in East Anglia and Lincolnshire receive an average annual rainfall total of less than 600 mm.

The main reason for the variation of rainfall in the British Isles is the **prevailing wind** direction. Most of the time the British Isles is affected by winds from the south-west. These winds bring warm and moist air from the Atlantic Ocean. When they reach the British Isles, they are forced to rise up and over the mountains (Figure 2). This leads to cooling, condensation and the formation of rain clouds. This explains why the mountains receive the highest rainfall totals. When the air transfers to the east, it is much drier, accounting for the lower rainfall totals and the so-called **rain shadow** effect.

> **Prevailing wind**: the most frequent, or common, wind direction.
>
> **Rain shadow**: an area or region behind a hill that has little rainfall because it is sheltered from rain-bearing winds.

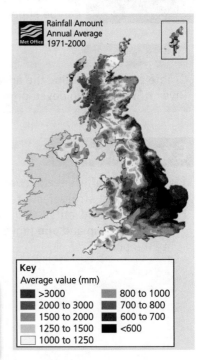

Figure 1 Rainfall map of the British Isles

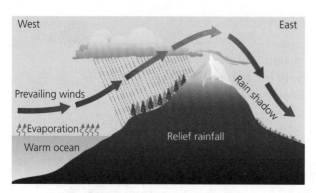

Figure 2 The formation of relief rainfall

Revision activity

Make a copy of Figure 2. Add detailed labels to describe the formation of relief rainfall.

Population density

REVISED

Population density varies a great deal across the British Isles (Figure 3).

- **Low population density**: much of northern Scotland has a low population density. This is because much of this region is mountainous and experiences a hostile climate. It is not an ideal location for human settlement and economic activity. There are also relatively low population densities in Northern Ireland and Wales.

- **High population density**: the rest of the UK, with its gently rolling hills and moderate climate, is much better suited for settlement. The highest density of population stretches from north-west England to the south-east. This area was developed as the UK's industrial heartland during the nineteenth century and attracted huge numbers of people to live and work here. Well served by transport routes, it remains the UK's most important economic region.

- **Very high population density**: with their high-rise housing blocks and modern housing estates, cities such as London experience extremely high population densities. Employment, shops and entertainment encourage many people to live in urban areas.

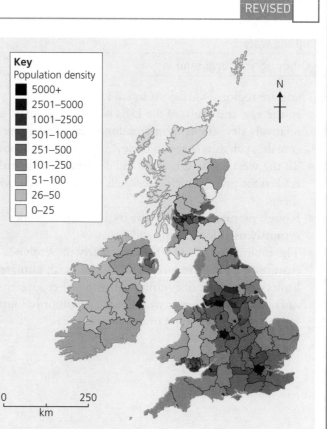

Key
Population density
- 5000+
- 2501–5000
- 1001–2500
- 501–1000
- 251–500
- 101–250
- 51–100
- 26–50
- 0–25

0   250
km

Figure 3 Population density in the British Isles, 2011

Land use

Figure 4 lists the land use in the UK. Notice that the vast majority of the UK is farmland and forest. Even within urban areas, nearly 80 per cent of the land use is 'natural' rather than 'built', comprising gardens, parkland and other green spaces.

Figure 4 Land uses in the UK

Land use	Percentage
Grasses and rough grazing	52
Arable/horticulture	20
Urban land use	14
Forest and woodland	12
Inland water	1
Other agricultural land	1

> **Revision activity**
>
> Select an appropriate method to represent the data in Figure 4.

There are regional variations in land use:

- In the east and south of the UK, where the climate is warm, sunny and relatively dry, arable farmland dominates. Farms specialise in growing cereals, such as wheat and barley, vegetables and root crops.
- To the west of the UK grassland dominates. The mild and wet climate is ideal for grass, which offers rich pastures for dairying, beef cattle and sheep.
- Rough pasture dominates the higher land in Wales and Scotland; this is mostly used for grazing sheep.
- The mountains of the UK, particularly in Scotland, tend to have rough pasture or heather moorland. Here the harsh climate and poor soils limits the growth of commercial crops. Sheep grazing is the main form of farming. Coniferous woodland is commonly found on poor acidic soils in relatively mountainous remote areas.

Water stress

In the UK, most rain falls in the west and north, whereas the greatest demand for water – for domestic use, industry and agriculture – is in the south and east.

This mismatch between supply and demand creates **water stress** in those areas where water is in limited supply but where there is a large and growing demand for it (Figure 5). Most areas with serious water stress are in the south and east, where the lowest rainfall totals occur. In the future, these areas may suffer from water shortages and even drought.

There are several possible solutions to prevent water stress:
- transfer water from the wetter west to the drier east; this could involve rivers, canals or pipelines
- construction of new reservoirs in the east to capture and store water; this would be very expensive
- encouraging water conservation by reducing leaks from pipes and encouraging people to use less water.

> **Water stress**: pressure on water supplies caused by demand exceeding or threatening to exceed supply.

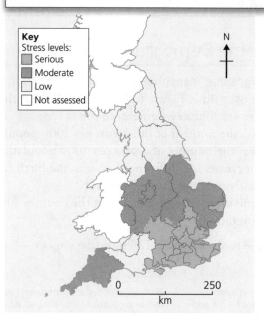

Key
Stress levels:
- Serious
- Moderate
- Low
- Not assessed

0   250
km

Figure 5 Water stress in England

Now test yourself

1 Use Figure 5 to identify the areas in England suffering from the most serious water stress.
2 How can the problem of water stress in these areas be addressed?

Housing shortages

The UK is facing a shortage of housing, particularly in London and the South East, where demand is greatest.
- Up to 250,000 new homes need to be built each year to keep pace with demand – currently just over 110,000 are being constructed annually.
- An estimated 160,000 new homes will be needed in the next five years in southern England to keep pace with demand.
- Some 15,000 new homes will be needed in London alone.

High-density flats or small houses are well suited to urban environments such as London, Birmingham and Manchester. These allow large numbers of people to be settled in a relatively small area of land.

A lot of new housing developments are taking place on the edges of towns and cities, however. These developments threaten the green belts – the zones of countryside with strict planning controls that surround most large towns and cities.

In 2014 the government announced plans to develop two 'garden cities', Ebbsfleet in Kent and Bicester in Oxfordshire. Some 15,000 new homes are planned for Ebbsfleet and 13,000 new homes are planned for the outskirts of Bicester. The homes will create village communities with plenty of green space and opportunities for shopping, recreation and employment. In 2017 the government announced an expansion of its garden city initiative.

Exam practice

1 With reference to Figure 1 on page 106, describe the pattern of rainfall in the UK. [4]
2 Describe a suitable presentation technique to illustrate the land-use data in Figure 4 on page 108. [2]
3 Explain how the characteristics of the UK can create water stress. [4]
4 To what extent do the physical and human characteristics of the UK contribute to the housing shortages? [6]

Population trends and the Demographic Transition Model (DTM)

What is the Demographic Transition Model (DTM)?

REVISED

The **Demographic Transition Model (DTM)** is a graph showing trends in population over time. Figure 6 shows the DTM for the UK from 1700 to 2020. There are four key elements to the DTM:

- birth rate: the number of live births per 1000 population per year
- death rate: the number of deaths per 1000 population per year
- natural increase: the difference between the birth rate and the death rate, usually expressed as a percentage
- total population: total population of the country (natural increase plus/minus migration).

It is possible to divide the DTM into five stages:

> **Demographic Transition Model (DTM):** a graph showing trends in population over time.

Stage 2

Improvements in living conditions, health care and diet, and the introduction of medicines, cause the death rate to fall. Birth rate is still high, causing the population to increase.

Stage 3

Birth rates fall because improved health care reduces infant mortality, so parents have fewer children. Women are now being educated, choosing to marry later in life and have fewer children. Death rates continue to fall and then level off. The total population continues to rise before starting to slow down.

Stage 4

Both the birth rate and the death rate are low. In an advanced society, infant mortality is low and women are choosing to follow careers, marrying later and having fewer children. Contraception is widely available and is used. Death rates reflect an **ageing population** so may start to rise slightly. The total population levels off.

Stage 1

Here both the birth rate and death rate are high, effectively balancing each other out. This explains why the total population remains largely unchanged. The high birth rate reflects high infant mortality (many are born because few survive) and the need for children to work and support the family. The high death rate is due to poor living conditions, lack of health care and widespread disease.

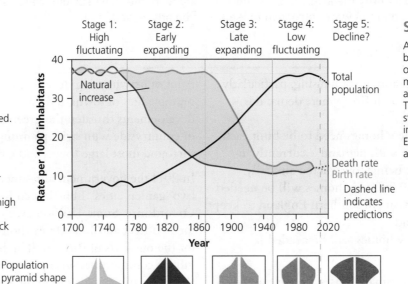

Stage 5

As the population becomes increasingly older, the death rate may start to creep above the birth rate. This has already started to happen in some Eastern European countries and in Germany.

Figure 6 The UK Demographic Transition Model (DTM), 1700–2020

Now test yourself

TESTED

1 Define the term 'natural increase'.
2 Name the two factors that determine the 'total population' of a country.

> **Ageing population:** a population structure that becomes distorted with a high and increasing proportion of people in middle and old age.

Revision activity

Draw your own sketch of the DTM and use labels to describe the reasons for the changing trends.

What are the trends in the UK's population since 2001?

The UK is in Stage 4 of the DTM. While the birth rate and death rate are both low, the birth rate still exceeds the death rate.

- The total population grew by some 4 million people between 2001 and 2011, and now stands at about 63 million, of which 84 per cent live in England and Wales.
- Natural increase together with immigration (particularly from Asia and Eastern Europe) accounts for this population rise. The number of immigrants settling in the UK rose from 4.6 million in 2001 to 7.5 million in 2011. Most immigrants are young, which in part explains the recently rising birth rate.

Despite the slight increase in the birth rate, the UK has an increasingly ageing population, with over 10 million pensioners (9.4 million in 2001; 10.4 million in 2011). As people live longer, there is a considerable strain on the health service.

> **Revision activity**
>
> Use Figure 7 to make a list of **three** changes that have occurred to the UK's population pyramid between 2001 and 2011.

Now test yourself

TESTED

Why is the UK in Stage 4 of the DTM?

The UK's population pyramid

REVISED

The structure of the UK's population – its breakdown by age and sex – can be shown by a graph called a **population pyramid** (Figure 7). Look back to Figure 6 to notice that it matches the overall shape of the population pyramid in Stage 4 of the DTM.

> **Population pyramid**: a diagram, essentially a bar graph that may resemble a pyramid shape, that shows the structure of a population by sex and age.

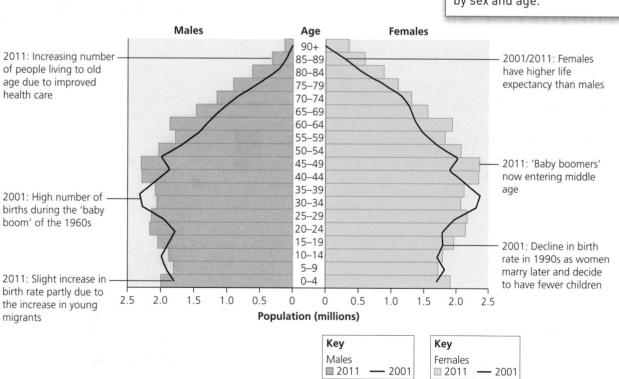

- 2011: Increasing number of people living to old age due to improved health care
- 2001: High number of births during the 'baby boom' of the 1960s
- 2011: Slight increase in birth rate partly due to the increase in young migrants
- 2001/2011: Females have higher life expectancy than males
- 2011: 'Baby boomers' now entering middle age
- 2001: Decline in birth rate in 1990s as women marry later and decide to have fewer children

Key Males ▪ 2011 — 2001

Key Females ▪ 2011 — 2001

Figure 7 Population pyramid for the UK, 2001 and 2011

The UK's ageing population

Causes of the UK's ageing population

In 2011, some 16 per cent of the UK's population was aged over 65. The UK's population pyramid (Figure 7) shows a notable 'bulge' of population (the so-called 'baby boomers' born after the Second World War) moving up the pyramid. With improved health care, better diets and improved lifestyles (less smoking and increased exercise), the UK is experiencing an increasingly ageing population.

Effects of an ageing population – challenges and opportunities

An ageing population presents both challenges and opportunities.

Figure 8 Challenges and opportunities presented by an ageing population

Challenges	Opportunities
Cost of health care will increase, placing pressure on the NHS and health care providers.	Elderly people can contribute to the economy by working part time or working from home and paying taxes.
Elderly people will require more support if they are to remain in their own homes.	Older people are often involved in voluntary work and in supporting other members of the family (child care, and so on).
Increasingly, middle-aged children who support their own children are now having to support elderly parents as well.	Some older people have considerable spending power, seeking leisure pursuits in retirement (cars, holiday homes). This can boost the economy.
Infrastructure (such as public transport, roads and pavements) needs to be modified to cater for elderly people.	Businesses providing services for older people (such as travel and medical services) can benefit.

Distribution of older people in the UK

Figure 9 shows the distribution of pensioners in the UK. While pensioners are distributed across the UK, there are some trends:

- relatively fewer pensioners in London and the South East due to the influx of large numbers of younger people of working age into the UK's economic hub
- high concentrations in the South West, a traditional retirement 'hotspot' with a warmer winter climate and gentler pace of life
- high concentrations at the coast (for example Sussex, Norfolk and Wales), reflecting people's desire to retire to the seaside to enjoy a slightly warmer climate and attractive landscape
- some areas have relatively high concentrations of older people, reflecting the out-migration of younger people; these are often remote rural areas such as western Scotland and parts of Wales.

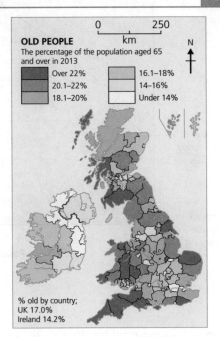

Figure 9 Distribution of pensioners in the UK

Now test yourself and exam practice answers at **www.hoddereducation.co.uk/myrevisionnotes**

Responses to the UK's ageing population

The government has responded to the UK's increasingly ageing population in a number of ways.

- Pensioner Bonds were issued in 2015 to encourage older people to save money at an improved rate of interest.
- Pensioners receive financial support, such as reduced transport costs and heating allowances (winter fuel payments). This support may be means-tested in the future, with only the poorest being eligible for support.
- Pension age has been increased – people have to work for longer before receiving their state pension (from 2018, both men and women will have to be 66 before receiving the state pension; between 2026 and 2028, it will rise to 67).
- Pro-natal policies, such as improved child care provision, are encouraging people to have more children to prevent the UK entering Stage 5 of the DTM.
- Allowing the immigration of young families to help boost the UK's birth rate.

Now test yourself

1 Why does the UK have an increasingly ageing population?
2 Identify **two** government strategies designed to address the issue of an ageing population.

TESTED

Revision activity

Draw a summary spider diagram to identify the challenges and opportunities of an ageing population. Can you add any points to those given in Figure 8?

Population change in a named place: Boston, Lincolnshire

Boston is a small market town in rural Lincolnshire in eastern England (Figure 10). It has a population of about 65,000.

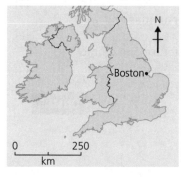

Figure 10 Location of Boston, Lincolnshire

How has the population changed in Boston?

REVISED

Between 2001 and 2011, the population of Boston grew by 15.9 per cent, compared to 10.4 per cent for Lincolnshire as a whole and 7.9 per cent for England. The main reason for the rapid population growth of Boston was immigration from parts of Eastern Europe.

Who were the immigrants?

- In 2001, there were some 200 non-UK born residents, mostly from Germany.
- In 2004 and 2008, several Eastern European countries joined the EU, allowing their citizens to move freely within the EU. Many people travelled to Lincolnshire to work on farms or in food processing factories, where they could earn more money than in their home countries.
- By 2011, ten per cent of the population of Boston came from Eastern European countries such as Poland, Lithuania and Latvia. Most immigrants are aged between 20 and 30.

How has the population change affected Boston?

REVISED

The recent influx of immigrants has created opportunities and challenges.

Figure 11 Challenges and opportunities presented by immigration

Challenges	Opportunities
Increased pressure on services, such as schools and health care, especially with an increasing number of children	Greater ethnic diversity, introducing new shops, foods and cultural traditions
Occasional social conflict between newcomers and the older resident population	Birth rate has increased, balancing the trend of an ageing population
Perceived loss of jobs for the resident population, as immigrants are willing to accept lower wages	Supporting the economic growth of the region by offering relatively cheap labour
	Support in care services and other trades

Now test yourself

TESTED

Why have large numbers of Eastern Europeans moved to Boston?

Exam practice

1 Explain the position of the UK on the Demographic Transition Model. [4]
2 Outline the causes of the UK's ageing population. [4]
3 Use Figure 9 to describe the distribution of the UK's pensioners. [4]
4 Suggest how the UK's increasingly ageing population can create both opportunities and challenges. [6]
5 With reference to a named place in the UK, explain how the population structure and ethnic diversity have changed since 2001. [6]

ONLINE

Exam tip

Remember that when 'explaining' (for example in Exam practice questions 1 and 5) you need to give reasons rather than just giving a description.

In question 5, ensure that your answer is balanced when writing about population structure and ethnic diversity. You must focus on a named place, using some specific relevant facts and figures.

Economic changes in the UK

The UK is one of the largest economies in the world. In the past, the UK's economy was dominated by heavy manufacturing industry based on a rich supply of natural resources.

Here are some facts about the UK's economy in the past:

- Most industries were powered by coal mined from south Wales, the Midlands, north east England and Scotland.
- Towns and cities grew up producing steel, ships and textiles. Ports such as London, Liverpool, Glasgow and Bristol developed as important hubs for imports and exports.
- In the last few decades of the twentieth century, many of the UK's industries became outdated and faced competition from abroad. Factories closed and many people lost their jobs.

The UK's economy is now dominated by the service sector. This includes financial services; high-technology industries based on research; media and creative industries; and tourism.

There is still some manufacturing in the UK, including cars, chemicals, light engineering and food processing. These modern industries are very efficient and can compete on the world market. Construction (for example housing, road and railway building) is important in the UK. For example, Crossrail will open in 2018, linking west and east London with Heathrow Airport.

The impact of recent government policies

Changes in the UK's economy have been driven by technological developments, world markets and government policies. Since 1997 the UK has experienced a Labour government (1997–2010), a Conservative/Liberal Democrat coalition government (2010–15) and a Conservative government (since 2015). In recent years the UK's economy has been influenced by the following factors:

- **1997–2008:** UK economy grew steadily, based on political decisions to keep taxes low and enable private companies to thrive.
- **2008:** the global financial crisis had a massive impact on the UK's economy and many others around the world. As the banking system started to collapse, the UK government had to step in to support UK banks and building societies.
- **2008–10:** the recession that followed caused unemployment to rise from 1.6 million in January 2008 to 2.5 million by October 2009.
- **2010–15:** the new coalition government introduced spending cuts to reduce the country's huge financial deficit.
- **2015 onwards:** the strong growth of jobs in the private sector reduced unemployment to 1.6 million (4.8 per cent) in September 2016, its lowest level since 2005. This happened despite the UK voting to leave the EU in June 2016 (Brexit).

How has employment changed since 2001?

Today the UK economy is dominated by the service sector. It accounts for nearly 80 per cent of the UK's GDP and employs about 80 per cent of the working population. The changes in employment between 2001 and 2011 are shown in Figure 12.

Figure 12 Changes in employment, 2001 and 2011

Employment sector	2001	2011
Agriculture	2%	1.5%
Industry (including construction)	28%	18.8%
Services	70%	79.7%
TOTAL	**100%**	**100%**

Now test yourself

1. How did the 2008 global recession affect the UK's political priorities?
2. Suggest the potential impacts of the UK's referendum decision to leave the EU.

Revision activity

Use the data in Figure 12 to draw two pie charts to show the UK's employment sectors in 2001 and 2013.

The changes can be summarised as follows:

● Agriculture has continued to decline slightly due to increased mechanisation and amalgamation of farms into larger units.
● Industry has declined, mainly due to mechanisation and competition from abroad.
● Services have grown due to expansion in financial and professional services, tourism, media, education and health care.

Figure 13 shows changes in some selected industries. It supports the findings in Figure 12.

Figure 13 Workplace employment for selected industries in the UK, 2001 and 2013

Employment sector	Employment (thousands)		
	2001	2013	Percentage change 2001–13
Agriculture, forestry and fishing	388	368	-5
Mining and quarrying	72	64	-10
Manufacturing	3,519	2,420	-31
Construction	1,955	2,019	3
Accommodation and food services	1,756	1,976	13
Professional, scientific and technical	1,713	2,621	53
Education	2,217	2,723	23
Arts, entertainment and recreation	752	890	18
TOTAL	**28,580**	**30,677**	**7**

The government has started a programme to create a more balanced economy by promoting the growth of manufacturing in the UK. This will focus particularly on the high-tech scientific and engineering sectors, together with green energy. As a result, Siemens opened a £310 million wind turbine blade factory in Hull in Yorkshire in 2016.

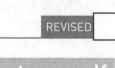

Now test yourself

Suggest why the service sector witnessed significant growth in the period 2001–13.

TESTED ☐

How have working hours changed since 2001?

REVISED ☐

Between 2001 and 2011 there were a number of changes in people's working hours. The main change was a slight reduction in average working hours of both men and women. The changes are summarised as follows:

● The average hours worked by men in full-time jobs fell from 46 to 44 hours a week.
● For women, average hours in full-time employment fell from 41 hours to 40 hours a week.
● Fathers worked shorter hours – in 2001, 40 per cent of fathers worked 48 hours or more a week. By 2011 this had fallen to 31 per cent.
● Fathers also worked less at weekends and in the evenings – in 2001, 67 per cent worked in the evenings, whereas in 2011 the figure had fallen to 50 per cent.
● The proportion of households with two full-time earners increased from 26 per cent to 29 per cent.

Now test yourself

Identify three trends in working hours during the period 2001–11.

TESTED ☐

UK economic hubs

The pattern of core UK economic hubs

REVISED

An **economic hub** is a centre of economic activity. As Figure 14 shows, economic hubs are widely distributed across the UK. They can occur at a variety of scales:

- The M40 in Oxfordshire has become a centre for motorsport industries.
- University towns, such as Oxford, Cambridge and Manchester, have tapped into their highly qualified graduate labour force to focus on the development of high-tech, research-based industries (biotechnology, software, medical and so on).
- Within cities, economic hubs have often developed as a result of deindustrialisation and urban redevelopment, for example Canary Wharf in London, Salford Quays in Manchester and the Titanic Quarter in Belfast.
- Throughout the country there are business and science parks located on the outskirts of towns and cities, often tapping into nearby university research facilities.

> **Economic hub**: a central point or area associated with economic success and innovation.

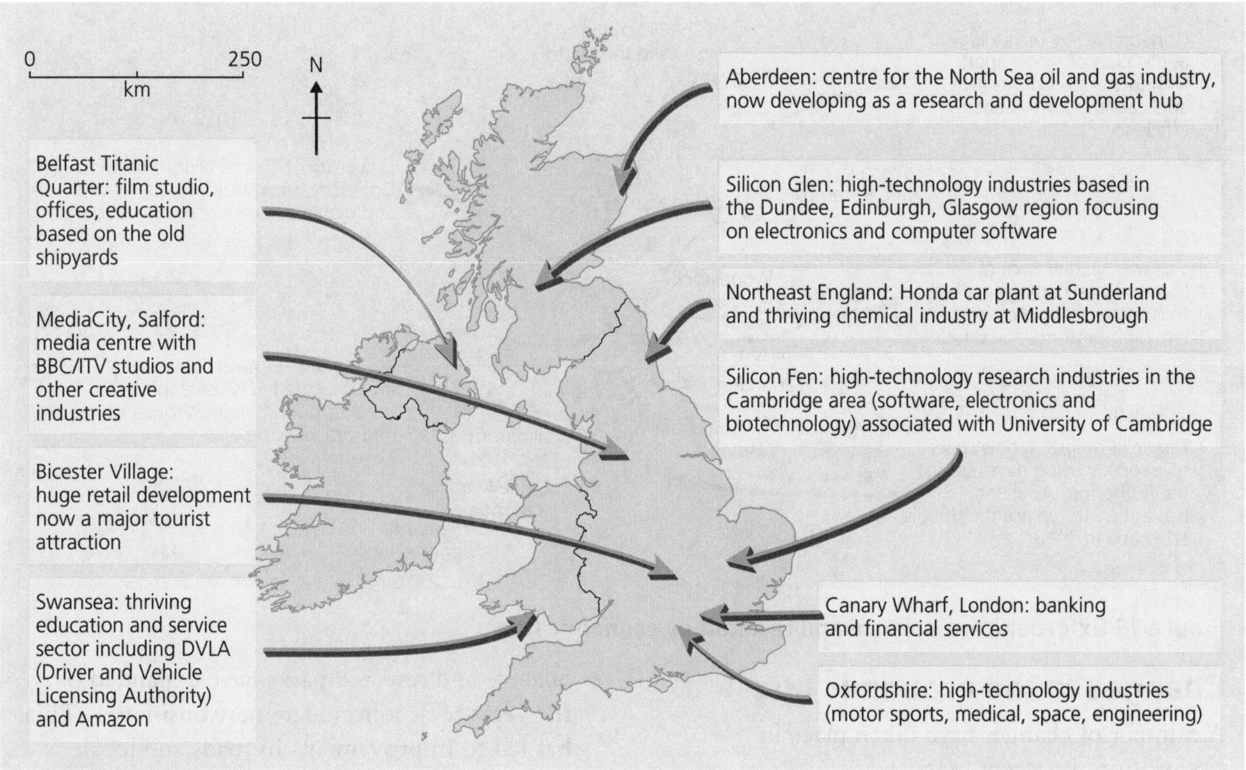

Belfast Titanic Quarter: film studio, offices, education based on the old shipyards

MediaCity, Salford: media centre with BBC/ITV studios and other creative industries

Bicester Village: huge retail development now a major tourist attraction

Swansea: thriving education and service sector including DVLA (Driver and Vehicle Licensing Authority) and Amazon

Aberdeen: centre for the North Sea oil and gas industry, now developing as a research and development hub

Silicon Glen: high-technology industries based in the Dundee, Edinburgh, Glasgow region focusing on electronics and computer software

Northeast England: Honda car plant at Sunderland and thriving chemical industry at Middlesbrough

Silicon Fen: high-technology research industries in the Cambridge area (software, electronics and biotechnology) associated with University of Cambridge

Canary Wharf, London: banking and financial services

Oxfordshire: high-technology industries (motor sports, medical, space, engineering)

Figure 14 Selected UK economic hubs

Exam tip

The specification requires you to study a named example of an economic hub. If you decide to study your own example, make sure that you focus on the changes and the regional/national impacts in your revision.

The changes in one economic hub

Case study: High-technology industries in Oxfordshire

Oxfordshire is one of the UK's most important locations for scientific and technological industries, including high-performance engineering, space and medical research. There are an estimated 1500 high-technology companies in the Oxford region (see Figure 15), varying in size from small start-ups to large multinationals such as Unileverw.

There are several reasons why Oxfordshire is an ideal location for the high-technology industry:

- The University of Oxford and Oxford Brookes University provide first-class research and teaching. Many graduates work in the high-technology industries.

- Oxfordshire has a rich history and beautiful countryside, making it an attractive location to live and work.

- Several large established research organisations, such as the UK Atomic Energy Authority and the Medical Research Council, have attracted new companies to the area.

- Oxford has a long tradition as a centre for the UK motor industry. Currently owned by BMW and producing Mini cars, Plant Oxford (as it is now known) in Cowley started production over 100 years ago.

- Oxfordshire is well served by road and rail. It is just 40 miles from Heathrow Airport and 50 miles from London.

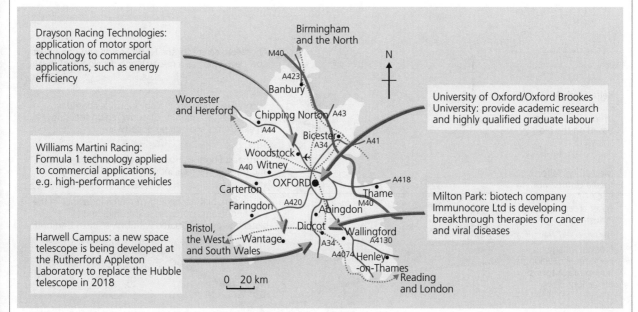

Figure 15 Oxfordshire: a science and technology economic hub

Changes to the economic hub

A number of changes have taken place in Oxfordshire in recent years:

- Building on Oxford's heritage of motor manufacturing, several Formula 1 companies have established research and development facilities in the region.

- The presence of large scientific and medical research organisations has encouraged similar companies to set up in the region. The clustering of companies brings many benefits, such as sharing ideas and new technologies.

- Science and research parks have expanded over the years to accommodate new businesses. This has led to improvements in roads and local infrastructure.

- Harwell Campus is planning several new developments, including a new space telescope and a residential quarter with 1400 new homes.

- Structurally, there have been several mergers of companies, particularly in the biotech sector. This reflects the rapidly changing and somewhat fragile nature of the scientific and technological industries.

Regional and national impact

The development of the hub in Oxfordshire has both regional and national impacts. They can be positive and negative.

Figure 16 Regional and national impact of Oxfordshire's high-tech economic hub

Regional impact	National impact
The high-tech companies employ 43,000 people – in the early 2000s, employment expanded by 40 per cent compared to a UK average of 18 per cent.	Oxfordshire's high-tech companies bring investment into the UK and boost revenue through paying taxes.
The rapid growth of these industries has put pressure on housing (Oxford has very high house prices), roads and other infrastructure. Commuting times are high and air pollution is an issue.	Modern research and manufacturing companies bring a certain amount of glamour and prestige to the UK (for example, Formula 1), attracting other investment from foreign companies.
A vast amount of money is spent in retailing, restaurants and hotels, supporting the local economy.	Several local industries have grown to become major international organisations; as an example, Oxford Instruments now employs over 1500 people across the world.
With strong international links, many people associated with high-tech companies choose to visit Oxford, boosting tourism.	The successful developments in Oxfordshire can have a positive economic impact on neighbouring regions.
Many local service industries (financial, legal, creative, and so on) have benefited from the growth of the high-tech sector. This includes construction, maintenance and servicing.	By increasing economic development in the 'south', the economic difference between the north and south may become greater.

1 What is an 'economic hub'?
2 Outline three characteristics of Oxfordshire's economic hub.
3 List three recent changes that have taken place in Oxfordshire's economic hub.

Revision activity

Draw a spider diagram to identify a selection of regional and national impacts of the growth of Oxfordshire's economic hub. Separate out advantages from disadvantages.

Exam tip

In Exam practice question 2, you must focus on your named UK economic hub (for example, Oxfordshire's high-tech industry). To 'examine its significance' you must try to be critical and consider both good and bad points (advantages and disadvantages). Make sure your answer is balanced by considering both regional and national impacts.

Exam practice

1 Explain the changes in employment sectors and working hours since 2001. [6]
2 With reference to a named UK economic hub, examine its significance to the region and the UK. [8]

ONLINE ☐

The UK's role in political conflict

As a member of several global organisations, such as the World Bank, the UN, NATO and the EU, the UK has a significant political role in the world. In resolving political conflict, the UK has been involved in direct military action (for example in Iraq and Afghanistan) as well as non-military support and co-operation with other nations.

A conflict zone is an area (usually a country or part of a country) where two or more groups of people have a serious disagreement that results in economic, social or military aggression.

Recent examples of conflict zones include Syria, Myanmar (Burma), Israel/Occupied Palestinian Territories, Afghanistan, South Sudan and Ukraine.

Now test yourself

TESTED ☐

1 What is a conflict zone?
2 What is the difference between a refugee and a displaced person?

> **Unilateral**: the involvement by a single country. For example, the UK spends over £4 billion on foreign aid to improve health care and education and provide disaster relief around the world.
>
> **Multilateral**: groups of countries working together to help individual countries. Examples include the UN, the World Bank and the EU.
>
> **Refugees**: people who have been forced to move to a different country, often in fear of their lives.
>
> **Internally displaced people**: people forced from their homes but who resettle elsewhere within their own country.

Case study: The UK's political role in one global conflict

Case study: Somalia

Location

Somalia is located in the far east of Africa (Figure 1). Its main geographical characteristics are:

● The northern part of Somalia is called Somaliland. Bordering the Gulf of Aden, much of this area is mountainous.
● The rest of Somalia borders the Indian Ocean. The land here is much flatter and most of it is dry semi-desert. This is an extremely hostile environment to live in.
● Somalia has a population of 10.8 million people (about one-sixth that of the UK).

Why is there a conflict in Somalia?

Formed in 1960, Somalia is a country that has been torn apart by war.

● Disputes with its neighbours soon broke out over land that was settled by the Somali people.
● In 1991 President Siad Barre was overthrown and the country was thrown into chaos. There was no government for two decades and fighting between rival warlords ravaged the country.

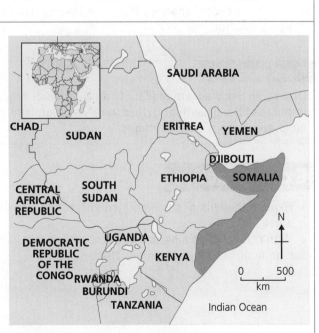

Figure 1 Location map of Somalia

● Droughts and famines in 1992 and 2010–12 killed an estimated 500,000 people in a country too poor to cope.

- In 2012 an internationally-backed government was installed and a degree of stability returned.
- The new authorities still face a challenge from rebels.

Within the country itself, there is unrest and disputes over land (Figure 2). In recent years Somalia has been linked to Islamic terrorism and also to piracy. Today Somalia is one of the most unstable places on Earth.

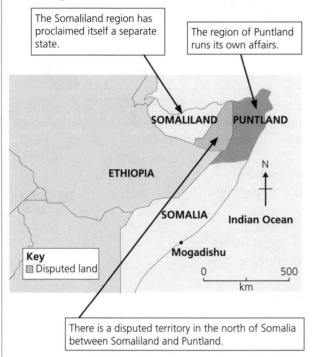

The Somaliland region has proclaimed itself a separate state.

The region of Puntland runs its own affairs.

There is a disputed territory in the north of Somalia between Somaliland and Puntland.

Figure 2 Disputed land in Somalia

What is the UK's involvement in Somalia?

The UK has had an important role in Somalia through its participation in international organisations such as the UN and the EU.

- The UK has long been a significant financial contributor to UN peacekeeping missions. In the early 1990s UN troops were sent to monitor a ceasefire between warring parties. This was largely unsuccessful and the troops withdrew in 1995, leaving the warlords to fight on.

- As a member of the EU, the UK supports EU initiatives such as the EU Naval Force (Operation Atalanta). Operation Atalanta seeks to protect vessels of the World Food Programme (WFP) and other vulnerable shipping from piracy in Somalian waters. It also supports other EU missions and international organisations that aim for security in the region.
- The UK's Department for International Development is working with the new Somali government to address shortages in food, employment and health care. Health care programmes, such as the Joint Health and Nutrition Programme for Somalia, aim to save thousands of lives by improving health services.
- In 2016, a British Army team was sent to Somalia as part of a UN mission to counter Islamic militants. The mission will support African Union peacekeeping efforts against the al-Shabab rebel group. About 70 UN personnel will eventually be in Somalia carrying out medical, logistical and engineering duties.
- In 2016, the UK government announced an investment of £1.5 million to improve infrastructure through the training of young people. This support will be delivered through the African Development Bank's (AfDB) Somalia Infrastructure Fund.

What were the impacts of the conflict?

- An estimated 500,000 people died as a result of the famines of 1992 and 2010–12, and tens of thousands of people fled to Kenya and Ethiopia in search of food.
- Every year thousands of Somalis attempt the perilous crossing of the Gulf of Aden to reach Yemen in search of security and a better life. Many die in their attempt.
- Many thousands have been displaced within Somalia. They live in overcrowded camps with poor access to food, water and sanitation.

Now test yourself

TESTED

Suggest why Somalia is one of the 'most unstable places on Earth'.

Exam practice

With reference to **one** named political conflict, examine the role of the UK through its participation in the work of international organisations. [6]

ONLINE

Revision activity

Create a timeline to record the main political events, conflicts and responses in Somalia from 1960 to 2016. Conduct internet research to bring this up to date.

The UK's media exports

The term 'cultural' is used to describe the values and beliefs of a particular society or group of people. It's all about what makes a society special. Culture is often illustrated by writings, paintings or creativity in the form of fashion, architecture and music.

The creative or media industries include films, television programmes, books, magazines, music, theatre and video games. The government estimates that the UK's creative industries are worth more than £63 billion a year, generating £70,000 every minute for the UK economy. They employ some 1.5 million people in the UK and account for £1 in every £10 of UK exports.

What is the global influence of the UK's media exports?

REVISED

- Since the London Design Festival started in 2003, over 80 cities around the world have started their own versions. This has stimulated home-grown creative designers and businesses, and they are now employing many people.
- UK architects have designed some of the world's leading sporting venues and facilities, for example the 2011 Cricket World Cup stadium in Pune, India.
- Approximately 100,000 students graduate annually in the UK in subjects such as architecture, design, music, fashion and digital media. Many are foreign students who return to their home countries to develop their own creative sectors.
- The UK is famous for its production of computer games, from Tomb Raider to Harry Potter to Grand Theft Auto. It has the largest games development sector in Europe, generating £2 billion in global sales each year. This creative output has stimulated the development of computer games in many other countries, such as the USA, Japan and Asia. Serious Games International is a British games company that has recently opened offices in Singapore employing local people to develop its products for foreign markets.

> **Now test yourself**
>
> Give three examples of the global influence of the UK's media exports.
>
> TESTED

Television programmes

REVISED

One of the most successful media exports is television programmes. The growth in international sales of UK television programmes has almost quadrupled since 2004. In 2013/14 it accounted for over £1.28 billion of export earnings.

British television programmes are popular abroad because of their originality and high-quality production. Sales of programmes include the original broadcasts together with rights to remake (format) the programmes for a local audience. *The Office*, for example, is shown in its original format in over 90 countries but has been licensed, for remakes in eight countries including Chile. This provides many opportunities for employment in the media sector involving both creative and technical work.

Among the most successful programmes to be exported in 2013/14 were *Atlantis*, *The Musketeers* and *Mr Selfridge*. Other popular exports include *Dr Who*, *Sherlock* and *Luther*.

The main markets are English-speaking countries, such as the USA (47 per cent of the market), Australia and New Zealand (Figure 3). However, the Chinese market is now expanding rapidly, increasing by 40 per cent from the previous year to £17 million. Two of the most popular exports to China are *Sherlock* and *Downton Abbey*. Other increasing markets are in South America, India and Scandinavia.

Other successful television exports include:

- *Who Wants to be a Millionaire* is Britain's most successful TV game show ever, with the format being exported to 107 countries including Indonesia, Algeria and Ecuador.
- *Top Gear* is said to be the most-watched factual TV show in the world. It has a global audience of some 350 million people in 170 countries.
- *Great British Bake Off* has been licenced by countries around the world including Austria, Ireland, Poland and Belgium.

Figure 3 Top ten countries for exports of UK television programmes, 2013/14

Country	Sales 2013/14 (£ million)	Change from previous year (%)
USA	523	+10
Australia/New Zealand	95	−10
Canada	75	−5
Sweden/Norway/Denmark	68	+8
France	37	+21
Italy	35	+13
Germany	31	+7
China	17	+40
Spain	16	−17
Netherlands	16	+28

Revision activity

Select an appropriate graph to represent the data in Figure 3.

Now test yourself

Outline the global influence of three British television exports.

TESTED ☐

Films

REVISED ☐

The UK film industry is a very important part of the UK's creative industries' exports. In 2013 it had a turnover of over £6 billion, accounting for 3.6 per cent of all UK creative industries. Figure 4 shows the main destinations of UK film exports.

- In 2013, the UK film industry exported £1.36 billion worth of film services, of which £756 million came from selling rights and £605 million from film production.
- In 2013, film exports were 73 per cent higher than when data was first collected in 1995.
- The James Bond blockbuster *Skyfall* accounted for £620 million alone.

- In 2013, 66,000 people were employed in the UK film industry, of whom 42,000 worked in film and video production. Employment is concentrated in London and the South East.

Most films produced today are international in terms of their funding, directing, acting and production, so it is not always possible to identify a completely 'British' film. The British film industry allows the UK to show off its strengths in acting, scriptwriting, production, music and visual effects. For example, in 2014 the film *Gravity* won the Oscar for Best Visual Effects for the London-based company Framestore.

Figure 4 Destination of UK film exports as a percentage of the total, 2009–13

Region	Percentage of total film exports
EU	41.5
Other Europe	6.5
USA	40.5
Asia	5.7
Rest of the World	5.8

Revision activity

Draw a pie chart to present the information in Figure 4.

Now test yourself

TESTED ☐

How does the UK film industry contribute to the UK's media exports?

Contribution of ethnic groups to the cultural life of the UK

The UK is a multicultural country. It has a long tradition of welcoming migrants from all over the world. Ethnic groups from many countries have chosen to migrate to the UK, some in search of employment and others forced from their homes by civil war or natural disasters. They have introduced aspects of their own cultures, such as music, food and fashion, and have contributed hugely to the economy and modern-day cultural landscape of the UK.

Case study: Ethnic food in the UK – Balti Triangle, Birmingham

Birmingham is one of the most multicultural cities in the UK. Some 30 per cent of its population come from ethnic minority groups, mainly from Pakistan, India and the Caribbean. The ethnic food retail sector is very important in Birmingham and one of the best-known areas is the so-called 'Balti Triangle' in the south of the city. Here a cluster of some 50 restaurants serve Balti dishes, a combination of meat and/or vegetables cooked with particular spices.

The dish was largely developed in Birmingham in the 1970s by people who had migrated from Kashmir, now a disputed region located within both Pakistan and India.

The Balti Triangle is an important tourist attraction. Apart from the restaurants, there are many shops offering cooking utensils, clothing such as saris and kaftans, jewellery, and Asian arts and crafts.

Exam practice

1 Assess the global influence of the UK's media exports, including television programmes and films. [6]
2 Through **one** example of food, media or fashion, describe its contribution to the cultural life of the UK. [4]

ONLINE

15 Will we run out of natural resources?

Our planet's natural resources – food, water and energy – are being overused and demand continues to rise. Consumption of resources is unevenly spread across the world. Some 80 per cent of resources are consumed by the 1 billion people in the world's richest countries. The USA consumes about a third of global resources, but less than five per cent of the world's population lives there.

Supply and demand of food, water and energy

Factors affecting supply and demand

Figure 1 Factors affecting supply and demand for natural resources

Resource	Factors affecting demand/supply
Food	Population growth, particularly in LIDCs – more people need more food
	Climate change – food supply has been impacted by unreliable rainfall, floods and droughts
	Land degradation – over-cultivation, overgrazing, soil erosion and desertification have reduced the land's capacity to grow food
	Changing food demands – demand for greater variety in produce has resulted in more commercial farming in LIDCs at the expense of growing staple foods; increased demand for meat in richer countries has resulted in crops being grown for animals rather than people.
	Trade – trade tends to favour the rich, leaving poor countries having to import basic foods
	Pests and diseases – sixteen per cent of the world's crops are lost to disease each year; pests thrive in wet, warm conditions and threaten new areas as the climate changes
Water	Population – population growth and changing lifestyles have dramatically increased the demand for water
	Industry – as countries become more developed, the demand for water in industry (for example, food processing) increases; industry currently uses about twenty per cent of the world's freshwater
	Agriculture – uses 70 per cent of the world's freshwater; demand for irrigation has increased, particularly in response to climate change
Energy	Use of fossil fuels – particularly coal, oil and natural gas – is increasing, especially in developing countries; these fossil fuels have a limited life and will begin to run out
	Improved lifestyles, industrial growth and the development of transport have led to a massive increase in energy demand; recent industrialisation in India and China has increased energy demand significantly
	Development of renewable resources has increased energy supply in certain regions
	Waste – over 25 per cent of energy is wasted each year or lost in production/transport; energy conservation is essential to preserve supply, but is limited in its adoption

Revision activity

Draw a spider diagram to outline a selection of factors responsible for increasing the demand for food, water and energy. Use colour coding to clarify the presentation of your diagram.

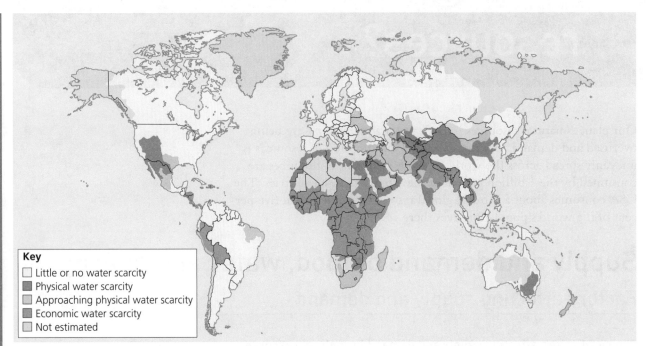

Key
☐ Little or no water scarcity
☐ Physical water scarcity
☐ Approaching physical water scarcity
☐ Economic water scarcity
☐ Not estimated

Figure 2 Many countries suffer from water scarcity

Now test yourself

TESTED ☐

1 'Resources are unevenly distributed across the world.' Is this statement true or false? Give one piece of supporting evidence.
2 Explain how population growth, increased consumption and changing technology increase the demand for resources.

How environments and ecosystems are modified by humans

In order to satisfy the increasing demand for resources, people have modified natural environments to increase production.

Mechanisation in farming and fishing

Farming practices have changed dramatically over time, largely driven by the move from subsistence to commercial farming. These changes have involved increased mechanisation, the expansion of irrigation, improved breeding of plant and animal species, and the widespread use of chemicals (fertilisers, pesticides and herbicides). Agricultural production has increased enormously and, as a result, famines are far less common than in the past.

Farming

The widespread use of tractors, combine harvesters, mechanical ploughs and motor transport has had an impact on environments and ecosystems.

- To accommodate large machines and improve productivity, small fields have been joined together by destroying hedgerows. This has had a significant effect on the environment as land is increasingly exposed to wind and rain, which leads to soil erosion.
- Hedgerows are important habitats for several animal species. Their removal has disrupted ecosystems and natural wildlife corridors for insects and small mammals.

- Mechanisation enables fields to be used throughout the year, which can lead to soils becoming exhausted. Field entries often become compacted (known as poaching) by heavy machines, leading to waterlogging, which affects soil ecosystems.
- Increased use of chemicals, together with vehicle fuels, can lead to pollution of water sources such as ponds and lakes, affecting those ecosystems.

Fishing

In commercial fishing, the use of mechanisation involves using large trawlers and huge nets extending several kilometres to sweep through the sea or across the seabed. Fish catches have increased dramatically since the 1950s when new technology was first introduced. Millions of people across the world depend on fish as a major source of food.

Commercial fishing has had harmful impacts on marine environments and ecosystems:

- Overfishing has threatened some species of fish, such as cod and tuna. Entire links have been removed from the aquatic food chain, upsetting the overall balance of the ecosystem.
- Some species – for example sharks, dolphins, turtles and seabirds – are caught up accidentally in nets and die. This also leads to ecosystem imbalance and, sometimes, the proliferation of prey animals such as rays.
- Fine-mesh nets catch smaller fish that are of little value to commercial fishing, reducing fish stocks and leading to future shortages.
- In tropical waters, commercial fishing often occurs close to coral reefs where there are high concentrations of valuable fish. Trawlers and nets can easily damage the coral, which unbalances the ecosystem, threatening the long-term future of the fish stocks.

Figure 3 Overfishing has threatened cod as a species

Now test yourself

1 Briefly outline how the following actions affect ecosystems:
   (a) removal of hedgerows
   (b) overfishing
   (c) use of fine-mesh nets.

TESTED

Deforestation and mining to provide energy

REVISED

Deforestation involves the removal of trees so that the land can be developed for other purposes, such as commercial farming, transport, settlements and mining. Vast swathes of tropical rainforest have been removed across large parts of South America and Asia, mainly to make way for commercial farming and mining.

Deforestation has occurred to enable TNCs to search for and extract valuable sources of energy such as coal, oil and natural gas. In Peru, oil and gas companies have removed trees in the Amazon rainforest to create access roads and drilling sites. Oil and gas developments have also resulted in deforestation in parts of the sub-Arctic, such as Siberia.

Impacts of deforestation

- Forests act as important carbon stores, absorbing carbon dioxide from the atmosphere. Deforestation – in particular, burning – releases the carbon into the atmosphere where it enhances the greenhouse effect, leading to global warming. Trees also help to regulate the world's water cycle by increasing atmospheric moisture.
- Trees provide habitats for many species of plants and animals; they form part of highly specialised ecosystems that may become severely damaged and unbalanced if deforestation occurs.

Impacts of mining for energy

- Opencast mining – often for coal – can cause enormous environmental destruction. Trees and plants are removed, destroying habitats and damaging ecosystems. Rivers and underground aquifers can be disrupted and polluted by the chemicals used in mining operations.
- Fracking – the extraction of oil and gas from shale rocks using high-pressure water and chemicals – has the potential to pollute underground water supplies and even trigger small earthquakes. Currently the USA extracts almost half of its oil and gas by fracking.
- When mining operations stop, landscapes can be left looking damaged and abandoned.
- Oil spills are extremely damaging to natural ecosystems. There have been several oil spills in Siberia (poorly maintained pipes) and the Niger Delta (sabotage). Oil tankers occasionally run aground, damaging marine ecosystems (for example *Exxon Valdez*, Alaska 1989).

Now test yourself

TESTED

1 Why is it so important to retain global forests?
2 What is fracking and why are some people concerned about its impact on the environment?

Reservoirs and water transfer schemes to provide water

Water supply can be increased and controlled by the construction of dams and reservoirs. An alternative option is to transfer water from areas of surplus to areas of deficit.

Reservoirs

The construction of a reservoir usually involves building a dam across a valley to hinder the flow of a river. The water ponds-up behind the dam and floods an area of land. These landscape modifications can have significant impacts:

- By interfering with the natural flow of a river, the aquatic ecosystem may be harmed, upsetting its natural balance.
- Habitats will be destroyed where land becomes flooded, as vegetation and soils become swamped and waterlogged.
- The natural landscape will be deeply transformed by the new constructions.
- Silt will collect behind the dam, reducing the capacity of the reservoir. In the long term this will have an impact on the newly developed aquatic ecosystem, for example by causing oxygen depletion.
- Natural floods will become less common, and seasonal wetlands downstream will dry up.

Water transfer schemes

A **water transfer scheme** moves water by canal or pipeline from a water source, such as a lake or a river, to an area of water scarcity and high demand. Often this involves transferring water from one river basin to another. Water transfer schemes can have environmental impacts:

- Invasive species – both plants and animals – can accidentally be transferred from one area to another, threatening and unbalancing ecosystems.
- Newly built pipes or canals can result in environmental damage and be harmful to local ecosystems.
- Water transfer can result in nutrient imbalances, which can affect aquatic plant and animal life. An increase in nutrients can promote excessive plant growth – especially of algae – deoxygenating the water through the process of **eutrophication**.

> **Water transfer scheme** moves water from a water source to an area of water scarcity and high demand.
>
> **Eutrophication**: the process of excessive nutrients (particularly nitrates and phosphates) building up in water sources, usually because of leaching and surface runoff.

Kielder Water and water transfer schemes

There are several examples of water transfer schemes in the UK. Kielder Water in Northumberland, for example, is the largest artificial lake in the UK. Construction began during the 1970s; it was completed in 1981 and filled the following year. Its original purpose was to boost water supplies for the heavy industries of the North East, in cities such as Newcastle. Ironically, deindustrialisation in the 1970s and 1980s meant that demand for water fell, and some criticised Kielder Water as being unnecessary.

Figure 4 Kielder Water

Kielder Water is a multi-purpose scheme: as well as storing water it generates electricity (hydroelectric power) and provides people with a range of leisure and recreational pursuits. Kielder Forest, a vast area of coniferous woodland planted around the edges of the reservoir, is the largest plantation in Europe.

The Kielder Transfer Scheme, costing some £167 million, transfers water from the North Tyne river to the industrial centres of Sunderland (via the River Wear) and Teesside (via the River Tees). The Kielder reservoir is spring fed, enabling water transfer to take place even during periods of prolonged drought.

Now test yourself

TESTED

1 Suggest **two** issues of environmental concern associated with:
   (a) constructing reservoirs
   (b) water transfer schemes.

Exam practice

1 Outline the factors leading to demand outstripping the supply of water. [4]
2 Suggest how the mechanisation of farming and commercial fishing to provide food can have harmful impacts on ecosystems. [6]
3 Examine the impact on the environment of deforestation and mining to provide energy. [6]

ONLINE

> **Revision activity**
>
> Use a table or diagram to summarise the potentially harmful impacts on the environment and on ecosystems of farming, fishing, deforestation, mining, reservoirs and water transfer schemes.

> **Exam tip**
>
> In Exam practice question 3 it is important that you make the connection between deforestation and mining, and their impacts on the environment.

Food security and the factors which influence it

What is food security?

Food security is achieved when people have physical and economic access to sufficient, safe and nutritious food that meets their dietary needs and enables them to maintain a healthy and active life. It is estimated that some 800 million people do not enjoy food security. Most of these people live in Africa, the Middle East and parts of Asia (Figure 1). While most of the world's richer countries enjoy food security, there are some individuals in these countries who do not.

> **Food security:** when people have physical and economic access to sufficient, safe and nutritious food that meets their dietary needs and enables them to maintain a healthy and active life.

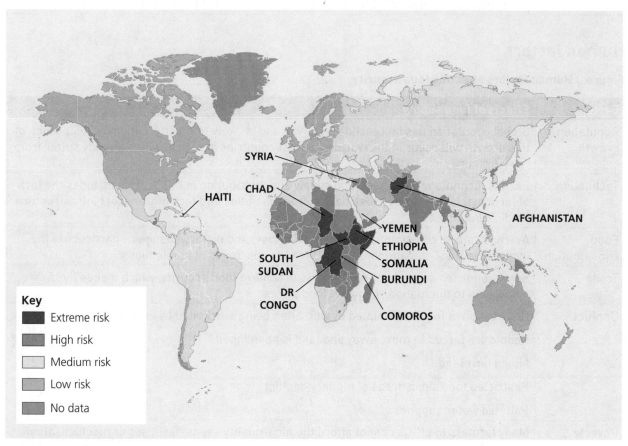

Figure 1 Global food security risk

Key
- Extreme risk
- High risk
- Medium risk
- Low risk
- No data

Look at Figure 1. Where are the areas of greatest food security? Which areas are at greatest risk from food insecurity?

What factors affect food security?

Several physical and human factors affect food security. Many of these are interconnected and affect mostly African LIDCs.

Physical factors

Figure 2 Physical factors affecting food security

Factor	Effects of food supply
Climate	Drought – results in severe food shortages and migration; can lead to desertification and salinisation. Floods – serious flooding results from tropical storms, which can devastate crops, for example Haiti and Fiji (2016). Climate change – patterns of rainfall appear to be changing. Some areas may become more productive whereas others may suffer from more frequent droughts and floods.
Pests and diseases	Many tropical regions suffer from pests and diseases that affect both animals and plants. In poor societies, people may also suffer from disease, reducing their capacity to be productive.
Water stress	Water stress is a serious issue in many LIDCs, particularly in Africa where climate change is expected to make the situation worse. The lack of water security and drought reduces food production. LIDCs cannot afford expensive water transfer schemes.

Human factors

Figure 3 Human factors affecting food security

Factor	Effects of food supply
Population growth	Global population has increased since 1950 and is likely to reach 9 billion by 2050. Much of this growth will occur in the world's poorest countries in Africa, which already suffer from food insecurity.
Technology	Lack of technology can affect food supply and distribution in LIDCs. This includes the lack of farm machinery (low yields), poor storage facilities and lack of transport infrastructure (to distribute food).
Food consumption	Average calorie consumption is expected to rise, and dietary changes – particularly the increase in meat consumption – could lead to shortages of staple foods.
Trade	Some countries depend on food imports to achieve food security, which makes them vulnerable to fluctuations in global food prices.
Conflict	Distribution of food is disrupted by war, often being stockpiled by armies. People are forced to move away and land is abandoned. Mined farmland. Restricted food aid in areas of military conflict. Polluted water supplies.
Poverty	Many farmers in LIDCs cannot afford the high-quality seeds, fertiliser or mechanisation required to produce more food.

Now test yourself

TESTED

Explain how climate and conflict can affect food supply.

Revision activity

Draw a spider diagram to summarise the main causes of food insecurity.

World patterns of food access

World patterns of food access can be illustrated by the World Hunger Index (WHI) and average daily calorie consumption.

The World Hunger Index

The WHI was developed by the UN as a means of examining progress towards the Millennium Goal 'to eradicate extreme poverty and hunger'. The index uses a range of indicators including undernourishment, the proportion of underweight children and the rate of child mortality. Each country is given a value from zero (no hunger) to 100 (extreme hunger).

As Figure 4 shows, the highest values (most extreme hunger) are in Africa. Parts of the Middle East and Asia also suffer from 'alarming' levels of hunger.

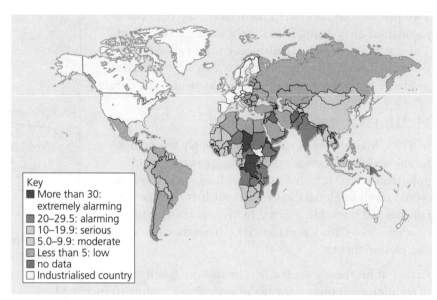

Key
- More than 30: extremely alarming
- 20–29.5: alarming
- 10–19.9: serious
- 5.0–9.9: moderate
- Less than 5: low
- no data
- Industrialised country

Figure 4 Global patterns of hunger according to the World Hunger Index, 2011

Average daily calorie consumption

Figure 5 is a choropleth map that shows the global average adult calorie consumption. On average, adults need about 2500 calories a day to live a healthy life. Notice the following patterns:
- Calorie intake is very uneven across the world.
- The highest calorie intake is in North America and parts of Europe. Obesity is a growing problem here.
- The lowest calorie values are found in Africa together with isolated countries in the Middle East and Asia. Many of those countries suffer from serious food shortages leading to malnourishment (lack of food/calories) and undernutrition (lack of a balanced diet). In the most serious cases, famine and starvation can result.

Remember that Figure 5 shows average values for countries as a whole. There can be huge variations in calorie intake within individual countries, both rich and poor.

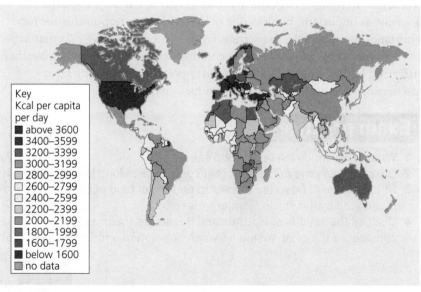

Key
Kcal per capita per day
- above 3600
- 3400–3599
- 3200–3399
- 3000–3199
- 2800–2999
- 2600–2799
- 2400–2599
- 2200–2399
- 2000–2199
- 1800–1999
- 1600–1799
- below 1600
- no data

Figure 5 Global patterns of average adult daily calorie consumption

Malthusian and Boserupian theories

Is there a relationship between population growth and food supply?

REVISED

Two people who considered the relationship between population growth and food supply were the Reverend Thomas Malthus and a Danish economist called Ester Boserup. Their opposing theories, which are summarised in Figure 6, have led to considerable discussion and debate.

Malthusian theory

In 1798, Malthus predicted that food supply would not be able to keep pace with the rapidly growing world population. He suggested that subsequent food shortages would result in global catastrophes, such as devastating famines and wars. He referred to historic famines in countries like China as evidence for his somewhat 'doom and gloom' theory.

Critics of his theory suggest that he had not taken account of technological improvements in agriculture leading to increased land productivity; neither had he considered that contraception and a reduction in child mortality would reduce the rate of population growth.

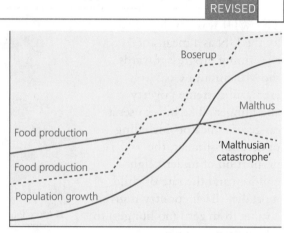

Figure 6 The theories of Malthus and Boserup relating food supply to population growth

Boserupian theory

Boserup suggested a more positive relationship between food supply and population growth in the 1960s. She argued that people would respond to future resource depletion by making technological advances that would lead to an increased food supply. She also argued that people can respond to shortages by reducing food consumption.

Globally we currently produce enough food to feed the world's population, but it is unevenly distributed. This explains the global variations in hunger and calorie intake (Figures 4 and 5). Population increase is important, but it is one of several factors responsible for food shortages. For example, the serious famine in China (1958–62) that killed 30 million people resulted from a combination of drought, poor weather affecting harvests and Communist Party policies prohibiting farm ownership (resulting in poor land productivity).

> **Now test yourself**
>
> Briefly outline the differences between the theories of Malthus and Boserup.
>
> TESTED

Exam practice

1 What is meant by the term 'food security'? [2]
2 Describe the physical factors that contribute to food insecurity. [4]
3 Study Figure 4. Describe the world pattern of food access as illustrated by the World Hunger Index. [4]
4 Discuss the validity of Malthus and Boserup's theories that considered the relationship between population and food supply. [6]

ONLINE

> **Exam tip**
>
> In Exam practice question 4, be prepared to offer your own opinion but make sure that you justify the points that you make. Remember the command word 'discuss' means that you should consider both sides of the argument.

Case study: Food security

Case study: Food security in the UK

The UK has a population of about 64 million people and it enjoys a high level of food security. Most people have access to sufficient, safe and nutritious food that allows them to maintain a healthy and active life.

Recent trends in food consumption

Calorie intake has actually decreased by about twenty per cent since the 1980s. Despite this decrease, the average weight of the population has increased. Levels of obesity in the UK have also risen. Much of this is linked to our increasingly sedentary lifestyle – we sit for most of our activities, including working and socialising – and lack of exercise.

Figure 7 shows some recent trends in food consumption. The main trends include:

- An increase in consumption of fresh fruit and fruit products, reflecting the expansion of supermarkets and increased imports enabling fruit to be available throughout the year.
- A reduction in potatoes, due to the availability of a wider range of vegetables, the availability of processed potatoes (chips) and alternatives such as rice and pasta.
- A reduction in sugar, in response to health concerns and dietary changes.

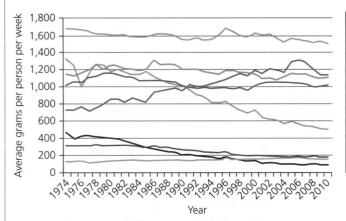

Figure 7 UK household food consumption, 1974–2010

Here are some facts about food imports and exports in the UK:

- The UK exported £18.9 billion worth of food and drink in 2013, £6 billion more than in 2005 (measured in 2013 prices).
- Imports of food and drink were much higher than exports, at £40.2 billion in 2013.

- In 2012, 53 per cent of the UK's food consumption was produced in the UK. The leading foreign suppliers to the UK were European countries, mainly the Netherlands (5.8 per cent), Spain (5.3 per cent), France (3.5 per cent), Germany (3.3 per cent) and Ireland (3.0 per cent).

Now test yourself

TESTED

1 Look at Figure 7. Describe the changes in UK consumption of fresh potatoes, and fresh fruit and fruit products.
2 Name three countries that export food to the UK.

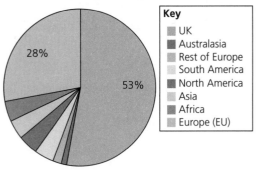

Figure 8 Origin of food consumed in UK, 2013

Achieving food security at the local scale

Figure 9 Strategies for achieving food security

Strategy	How food security can be achieved
Food banks	Food banks are stores of food that can be accessed by people in need.
	Held in central locations, food banks are supported by food donated by the public and organisations such as schools, churches and local businesses.
	Doctors, health visitors and social workers provide vouchers to people in need, entitling them to three days' worth of food.
	An estimated 1 million people are dependent on over 400 food banks in the UK.
	Despite low rates of unemployment, many people do not earn enough money to support themselves and their families.
Urban gardens	Urban gardens are productive plots of land in towns and cities that are used by individuals or community groups to grow food (vegetables, fruit, herbs, and so on).
	Some schemes make use of ground that was previously wasteland.
	Community-run gardens engage local people and help to encourage healthy eating.
	One example is the Incredible Edible scheme in the West Yorkshire town of Todmorden. Crops have been planted on roundabouts and road verges. Colleges and local businesses have allowed herb gardens, orchards and raised vegetable beds to be planted on their land.
Allotments	Allotments are plots of land in towns and cities that are used for non-commercial crop production.
	Commonly administered by local councils or community groups, individuals pay a small annual rent to cultivate a small plot of land. Allotments are largely protected from development.
	The first allotments in the UK were created in 1809. They became very popular during the twentieth century, particularly during times of food shortages (for example, during the two World Wars).
	Currently there are about 300,000 allotments. While this is far fewer than in the past, recent concerns about local food security, food miles and global warming have led to a surge in popularity, particularly among younger people.

Now test yourself

TESTED

For **one** of the strategies in Figure 9, outline how it can improve food security at the local scale.

Revision activity

For your **one** chosen strategy aimed at improving food security at the local scale, create a spider diagram to outline the ways in which it works.

Exam tip

The specification requires you to study **one** attempt to help achieve food security at the local scale, such as food banks, urban gardens and allotments.

Past attempt: Green Revolution

The **Green Revolution** of the 1960s applied science to agriculture. The primary aim was to increase the production of basic food crops such as rice and wheat in the poorest countries of the world. New strains of seed known as high-yielding varieties (HYVs) were developed. An example is IR8, a 'miracle' type of rice that had the potential to be ten times more productive than traditional varieties.

While the Green Revolution boosted food production enormously in countries such as India and the Philippines, it did create some issues. To achieve the high yields, the new crops needed irrigation and the use of expensive chemicals, particularly fertiliser. Wealthy landowners achieved considerable success whereas the bulk of smallholders and peasant farmers benefited little; critics suggest that the gap between rich and poor increased as a result of the Green Revolution.

Other problems included:
● The loss of agricultural labouring jobs due to the introduction of machinery.
● Machines were expensive to maintain and fuel – this resulted in many tractors simply being abandoned.
● Pumps used for irrigation caused water tables to fall in some areas. This affected the flow in rivers and harmed ecosystems.
● The high productivity resulted in falling food prices, which reduced the income of smallholders.

Present attempt: GM crops

Genetic modification (GM) has, to some extent, been practised for a very long time. Most breeds of livestock and plants have been genetically modified by selective breeding. Domestic cattle, for example, have been selectively bred to maximise milk yields or meat production.

Modern genetic modification involves the introduction of changes in DNA through a process known as genetic engineering. This takes place in a science laboratory and is quite different from selective breeding.

GM crops can be bred to increase productivity as well as resistance to climatic extremes (such as drought), pests and diseases. The focus so far has been on commercial crops such as soybean, maize, oilseed rape and cotton seed. Soybean and maize are used widely as animal feed to produce meat.

The practice is widespread in the USA, Brazil and Argentina but has yet to be introduced in the UK. There are concerns about possible impacts on human health, the natural gene pool and on ecosystems.

> **Exam tip**
>
> The specification requires you to study **one past** and **one present** attempt to help achieve food security at the national scale such as global food trade, the 'Green Revolution', genetically modified (GM) crops and food production methods.

> **Green Revolution**: refers originally to the work of Norman Borlaug in India; the aim was to increase crop yields in LIDCs.
>
> **Genetic modification (GM)**: when genetic material (DNA) is altered in a laboratory rather than through selective breeding.

> **Now test yourself**
>
> 1 What is the 'Green Revolution'?
> 2 Assess the successes and failures of the Green Revolution.
> 3 How is selective breeding different from genetic modification (GM)?
> 4 Suggest why GM crops are not grown in the UK.
>
> TESTED

> **Exam tip**
>
> With Exam practice questions 1 and 2, be sure to select the appropriate example to write about. Make sure that you address the command words ('evaluate' and 'discuss') by considering both sides of the argument. You can include your own views.

Exam practice

1 Evaluate the success of **one** attempt to increase food security at the local scale. [6]
2 With reference to **one past** attempt, discuss its effectiveness in increasing food security at the national scale. [6]

ONLINE

Sustainable strategies to achieve food security

There are many possible strategies to achieve food security. However, to ensure long-term food security, strategies need to meet certain criteria:

- **Environmental sustainability:** involves no long-term damage to the environment or any harmful effects on ecosystems.

- **Economic sustainability:** ensures economic and job security, providing sufficient wealth for the community.
- **Social sustainability:** promotes community cohesion and support mechanisms for vulnerable people.

Ethical consumerism

Ethical consumerism involves the choices made by customers to buy particular products. For example, many people choose to buy products that are grown locally to support local businesses and reduce food miles. Some people make choices based on the welfare of animals, such as buying free-range eggs, or they may wish to buy organic produce.

Fairtrade

Fairtrade is a global movement that began in 1988 by guaranteeing Mexican farmers a fair price for their coffee. Today there are over 1.4 million farmers from over 70 countries whose produce is sold with the Fairtrade label. Fairtrade pays farmers a higher price and supports local community projects.

An important aspect of Fairtrade is the social premium that is paid to local producers, who are often organised into co-operatives (groups of farms sharing benefits). The social premium funds community improvements, such as water supply and education. This is a good example of social sustainability.

> **Ethical consumerism**: purchasing products or services produced in a way that minimises social and/or environmental damage.

Food waste

According to the UN, one-third of all food produced globally is lost or wasted. Food loss or wastage occurs from field to plate. In developed countries, most food waste occurs at the consumer stage, with people refusing to buy or eat misshapen fruit and vegetables or simply overbuying food and then not using it.

There is much controversy regarding 'best before' and 'sell by' dates. Food is often discarded by supermarkets when it is still perfectly safe to eat. Increasingly this surplus food is made available to charities and food banks. While this is a good thing in some respects, it is hardly sustainable as it does not solve the original problem. Some food waste is fed to animals or used as a biofuel to generate electricity. People are increasingly encouraged to use leftover food creatively, and several well-known chefs have published books on the subject.

Feeding Manchester is a network of social enterprises, charities and activists who are working to build a sustainable food movement in the city. The movement encourages people to buy, grow, sell and promote food that is local and seasonal, comes from sustainable farms and has reduced waste and packaging.

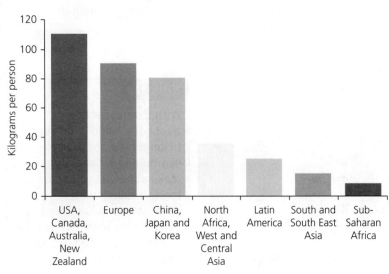

Figure 10 Annual food waste

> ### Now test yourself
>
> 1 What is Fairtrade?
> 2 How does the social premium benefit Fairtrade farmers and their communities?
> 3 Why is so much food wasted in developed countries?
>
> TESTED

Food production

Long-term food production is dependent on a sustainable approach to maintain soil fertility and water supplies. While some methods of food production use sustainable practises (for example, organic farming), some do not (for example, intensive farming).

Organic farming

Organic farming involves the production of food without the use of chemicals. It is widespread across the world, in both ACs and LIDCs.

Organic food production is in harmony with nature and reduces the harmful effects of chemicals. However, it tends to be more labour intensive (for example, weeding by hand) and yields are usually lower than for more intensive forms of farming. This means that the produce can be more expensive to the consumer and raises questions about its economic sustainability.

> **Organic farming**: the production of food without the use of chemicals.
>
> **Intensive farming**: maximising the yield from the land, often through mechanisation and the use of chemicals.

Intensive farming

Intensive farming involves maximising the yield from the land. It is often associated with mechanisation and the use of chemicals. It has some clear disadvantages:

- Monoculture involves growing the same crop in a field year after year, using chemical fertilisers and pesticides to raise productivity. Monocultures can damage soil structure and increase the risk of pests and diseases.
- Chemicals used in food production can leach through the soil and contaminate groundwater supplies.
- Excessive irrigation can cause water tables to fall – if abstraction exceeds recharge, groundwater supplies will be unsustainable.
- Chemicals could harm farm workers, affecting their economic and social well-being.

Now test yourself

TESTED

1 What are the main characteristics of organic farming?
2 In what ways is intensive farming unsustainable?

Technological developments

Technological developments make use of scientific research to promote crop growth. Such developments can incorporate sustainable principles. Alternatively, in maximising production they can have harmful impacts on the environment and be unsustainable.

GM crops

GM crops have been developed in some parts of the world to increase productivity (see page 137). However, often requiring irrigation and the use of chemicals, GM crops can be unsustainable, both economically and environmentally. There are also concerns about the possible long-term effects of genetic modification on the health of people and on natural ecosystems.

Hydroponics

Soil provides plants with the nutrients (plant foods) necessary for growth. New technology enables plants to be grown in a soil-free medium, with nutrients being delivered through an aerosol (aeroponics) or water (**hydroponics**).

Hydroponics is used to cultivate both fruit and vegetables. Closely monitored and controlled by scientists, plants can be grown throughout the year in artificially lit and heated buildings or greenhouses. The plants grow quickly as nutrients are applied directly to the roots, and diseases found in soils are eliminated. The plants can also be stacked as long as they receive enough light. This allows high levels of production.

Hydroponics can be environmentally sustainable as long as there is plenty of water and care is taken not to pollute the environment with nutrient-enriched water. It is, however, an expensive option. It requires a great deal of expert knowledge and may not be considered to be economically sustainable.

> **Hydroponics:** plants grown in a soil-free medium, with nutrients being delivered through water.

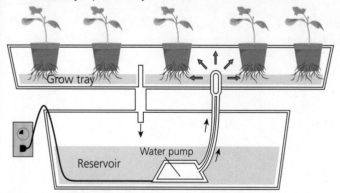

Grow tray is periodically flooded with the nutrient solution

Figure 11 Hydroponics

Now test yourself

1 What is hydroponics?
2 What are the advantages and disadvantages of hydroponics?

Bottom-up approaches

Bottom-up approaches are grassroots initiatives that come from individuals or community groups, as opposed to top-down approaches, which often come from governments or big business. The advantage of bottom-up approaches is that they address a specific need as identified by local people in a small area. They are often environmentally, economically and socially sustainable.

Urban gardens

Urban gardens are large, formal projects involving the production, processing and distribution of food from within urban areas. They often take the form of community initiatives involving the conversion of waste or derelict land into productive farmland or vegetable gardens. Throughout the world, people living in urban areas grow food on rooftops, in patio gardens and in back yards.

Examples of urban gardens include:
- **Michigan Urban Farming Initiative, Detroit, USA:** derelict wasteland has been converted into garden beds, providing employment opportunities and fresh food for local people.
- **InterContinental New York Barclay Hotel, New York, USA:** the hotel has an apiary on its roof; the honey produced by the bees is used in the hotel's kitchen.

The benefits of urban gardens include increased food security, healthier diets, improved natural environments (with birds, butterflies and bees) and greater social cohesion. Such initiatives are often economically, environmentally and socially sustainable.

> **Urban gardens**: large, formal projects involving the production, processing and distribution of food from within urban areas.

Figure 12 An urban garden in Queens, New York

Permaculture

Permaculture (*perm*anent agri*culture*) is similar to organic farming in that it promotes farming that is in harmony with the natural environment. Crops are planted and soils managed so that they can be used indefinitely.

Permaculture aims to reflect the way that natural systems operate without human interference. It uses natural systems rather than artificial chemicals, for example using ladybirds as natural predators for aphid control (aphids can decimate green crops) rather than chemicals. Other principles include producing zero waste, storing energy and using renewable resources.

Permaculture promotes a sustainable lifestyle involving practices such as rainwater harvesting, composting, crop rotation and woodland management.

Permaculture: crops are planted and soils managed so that they can be used indefinitely.

Revision activity

Create a spider diagram or table to summarise the main characteristics of the different approaches to achieving food security. Make sure that you refer to the environmental, economic and social sustainability of each approach.

Now test yourself

TESTED

1 What is the difference between top-down and bottom-up approaches?
2 Describe one example of an urban garden.
3 Define 'permaculture'.

Exam practice

1 Describe the food production technique of hydroponics. [4]
2 Explain why bottom-up approaches to food security are often sustainable and successful. [6]
3 Evaluate the sustainability of ethical consumerism in achieving food security. [6]

ONLINE

Exam tip

In Exam practice question 3 you need to focus on sustainability – try to refer to economic, environmental and social sustainability. In evaluating, consider good points and bad points, advantages and disadvantages. You could substitute 'ethical consumerism' for other strategies – such as 'food production', 'technological developments' and 'bottom-up approaches' – to give you additional exam practice.

17 Geography fieldwork

Assessment outline

As part of your GCSE course you will have completed two geographical enquiries, one physical and one human. Remember that in at least one of the enquiries you must show an understanding of both physical and human geography and their interactions. Make sure that you know which of your investigations involves this.

You will be expected to:
- apply knowledge and understanding to interpret, analyse and evaluate information and issues related to geographical enquiry
- select, adapt and use a variety of skills and techniques to investigate questions and issues, and communicate findings in relation to geographical enquiry.

Exam assessment
- The Physical Geography Fieldwork questions will appear in **Paper 1, Section B**.
- The Human Geography Fieldwork questions will appear in **Paper 2, Section B**.

It is possible to identify two distinct sets of questions about fieldwork in each of these sections.
- **Generic fieldwork questions**: these are based on the use of fieldwork materials from an unfamiliar context (such as graphs and diagrams) for you to criticise, or, for example, information about a location or sample strategy for you to evaluate.
- **Individual fieldwork enquiry questions**: these questions are based on **your** two enquiries. The questions will focus on reasons and justification, so you must understand **why** you did things and not just what you did.

Typical questions about individual fieldwork enquiries
- Explain how the theory behind the investigation determined the data-collection method(s) used.
- Evaluate the effectiveness of your data-collection methods.
- Justify the sampling strategies used in your enquiry.
- Assess the appropriateness of your data-presentation methods.
- Justify the choice of location(s) used to collect data.
- To what extent can the fieldwork results be deemed to be reliable?
- Assess the appropriateness of your data-collection methods.
- Evaluate the accuracy and reliability of your results/conclusions.

> **Exam tip**
>
> When preparing for the fieldwork questions in your exams, make sure you:
> - understand the reasons why you did things – the choice of location, sampling methods, presentation techniques
> - understand the higher-level command words, such as 'justify', 'evaluate', 'discuss' and 'to what extent'
> - revise your fieldwork enquiry notes – consider making revision notes
> - practise writing concise and precise answers.

18 Geographical exploration

Assessment outline

The Unit 3 Geographical Exploration exam paper will focus on a specific country, or a place within the UK, and will focus on at least two topics within Unit 1 and Unit 2.

Resource booklet

A separate twelve-page resource booklet will be provided alongside your exam paper. The resource booklet will contain a variety of geographical information linked to a particular theme. Information could include maps at different scales, diagrams, graphs, statistics, photographs, satellite images, sketches, extracts from published materials and quotes from different interest groups.

Exam assessment

The exam will consist of a series of questions that run through a geographical exploration of a specific country, or a place within the UK. You'll use the resources in the booklet to answer questions on different elements of the topic, and these questions will lead to a more extended piece of writing that will involve an evaluative judgement.

- **Early questions**: these will require you to interpret and analyse the information provided in the resource booklet.
- **Synoptic questions**: these will be longer 12-mark questions, which will require you to apply your knowledge and understanding of content from different topics that you have studied.
- **Final decision-making question**: this final question will require you to apply your knowledge and understanding of the information in the resource booklet to develop a critical perspective on the issue(s) studied. You will need to consider the points of view of the stakeholders involved, make an appraisal of the advantages and disadvantages, and evaluate the alternatives. You may, for example, be asked to suggest an appropriate project for urban improvement or to identify a preferred new road or rail route in the UK. You will be expected to justify your decision using the evidence available.

> **Exam tip**
>
> When studying the resources booklet and answering Paper 3, make sure you:
> - read through the extracts of text carefully
> - take time to study the photographs thoroughly
> - add notes and annotations to the booklet as you read through it
> - look for links and connections between the resources.
> - consider arguments for and against, or advantages and disadvantages
> - write concise and precise answers – don't waffle.

> **Exam tip**
>
> For the decision-making exercise, remember that there is no wrong answer, just one that is poorly supported by the evidence.